How to Build a Solar Heater

A complete guide to building
and buying solar panels,
water heaters, pool heaters,
barbecues, and power plants

TED LUCAS

A MENTOR BOOK

NEW AMERICAN LIBRARY

TIMES MIRROR

NEW YORK AND SCARBOROUGH, ONTARIO
THE NEW ENGLISH LIBRARY LIMITED, LONDON

This is an authorized reprint of an edition published by
The Ward Ritchie Press.

Library of Congress Catalog Card Number: 75-18097

MENTOR TRADEMARK REG. U.S. PAT. OFF. AND FOREIGN COUNTRIES
REGISTERED TRADEMARK—MARCA REGISTRADA
HECHO EN CHICAGO, U.S.A.

SIGNET, SIGNET CLASSICS, MENTOR, PLUME AND MERIDIAN BOOKS
are published *in the United States* by
The New American Library, Inc.,
1301 Avenue of the Americas, New York, New York 10019,
in Canada by The New American Library of Canada Limited,
81 Mack Avenue, Scarborough, Ontario M1L 1M8,
in the United Kingdom by The New English Library Limited,
Barnard's Inn, Holborn, London, E.C. 1, England.

FIRST MENTOR PRINTING, JANUARY, 1978

2 3 4 5 6 7 8 9 10

PRINTED IN THE UNITED STATES OF AMERICA

SOLAR POWER— THE ENERGY REVOLUTION!

With the dramatic increase in utility costs, the threatened decreases in service, and the government priority being given to the energy crisis, everyone has suddenly become aware of the energy and money-saving potential of solar heating. Yet very little practical information is available to the general public on this subject, even though so much of solar technology could easily be utilized by individual consumers. Now, this all-inclusive guide explains not only how to build solar heaters yourself but what they are, how they work, and how to select the right devices for your particular needs. Why spend the rest of your life paying someone else for energy. The sun's power is there for the using, and *HOW TO BUILD A SOLAR HEATER* can show you the way to make solar energy work for you!

"A practical, understandable approach to using solar energy . . . I have recommended it to my students."

—scientist Robert J. Schlesinger
of the California Institute of Technology

Ted Lucas earned his degree in physics from M.I.T. and is presently Senior Technical Writer for The Singer Company in Los Angeles. Recently he served as solar consultant for the Southern Pacific Development Corporation, setting up a system of solar power plants in Tahiti. In addition, Mr. Lucas has written many articles in leading technical journals, as well as a book, *Radar on Wings*, two volumes of poetry, and numerous short stories.

SIGNET and MENTOR Books of Special Interest

To my patient wife, Joan

Contents

Illustrations

Figures

Page

Figures   Page

Tables

1 / Savings from the Sun

THE ENERGY CRISIS

Everyone realizes that the world's supply of fossil fuels is limited. We are painfully aware that the cost of these fuels keeps rising by huge amounts, not only the gasoline for our cars, but also our utility bills for gas and electricity and, in many states, the oil or coal for heating.

There's no question that most of us would like to conserve our resources and reduce our fuel costs. When you consider that about 20% of U.S. energy consumption goes for space heating of buildings, another 4% for heating domestic water, and about 3.5% for air conditioning or space cooling, you realize that more than one-fourth of our fuel is lost forever every day—up various pipes, flues and chimneys.

A recent editorial in the *Los Angeles Times* bemoaned the fact that electric utility rates in Southern California have risen more than 100% in the last four years. Even more drastic rate increases have occurred in other areas. The writer pointed out: "What upsets a lot of people is that, even while they are paying more for energy, they are using less, and that seems to defy economic sense. . . . A Congressional report says utility bills for Americans were up $10 billion in 1974."

There is, however, an end to this inflationary spiral. Solar energy can provide you with more comfortable living at lower cost. This chapter will introduce you to the many applications of solar energy. By using the information in this book, you'll be able to save on fuel bills, whether you own a home, a cabin, a trailer, an apartment house, a commercial building, or a swimming pool.

If you are a "do-it-yourself" person, you can apply solar energy to heat *and cool* your home, to heat your water for domestic uses or for a pool. The instructions in this book will enable you to build any or all of these systems using ordinary hand tools and supplies found in neighborhood hardware and plumbing supply stores or conveniently provided in manufacturers' kits.

1

Help from Manufacturers

As with any relatively new industry, the people involved in the growing business of solar energy are eager to help anyone interested in using sunlight efficiently.

You can decide to do as much or as little as you wish in applying the benefits of solar energy to your life style. For a truly handy person, there is nothing difficult about building the kind of solar heater best suited to his or her specific needs. You may want to do all the work yourself, or with the help of friends. In addition to the detailed instructions in this book, there are references to many other sources of useful data.

If you prefer an intermediate route, you'll find in subsequent chapters suggestions on how to select solar collector panels and associated hardware from some thirty experienced manufacturers of this equipment. You can buy some of these systems in kit form, typically for heating water in your home or swimming pool. Other manufacturers offer you solar panels and instruction manuals with a list of suggested hardware ranging from plumbing, heat pumps and different thermostats to the nuts and bolts needed for installation.

Even if you want the work on your home, apartment building or swimming pool done by a contractor, you'll learn what is available and how to make intelligent choices. For instance, in Chapters 2, 4 and 5 there is a detailed analysis of the advantages and disadvantages of the two major types of solar home heating systems, air-type and liquid-type. In Chapter 3 you'll find a description of many varieties of solar collector panels—the structures which actually absorb the sun's heat—as well as how to build your own if you want.

SOME PRACTICAL USES OF SOLAR POWER

There are several ways to use solar power. On a world-wide basis, there are at least 500,000 persons using this free energy source, most of them using it to heat water for home use.

Heating Your Swimming Pool

If you own a swimming pool, you certainly want to consider adding a solar heating system. Year-round heating adds greatly to your enjoyment. In Chapter 8 you'll see how easy it is, whether or not you're experienced in doing your own home improvements, to use sunlight for heating your pool. If you're installing a new pool, or considering it, it's becoming mandatory in many areas to use solar energy. New York, Florida and California are among those having state laws or local ordinances requiring solar heating for all new outdoor pools.

Your cost for putting in a solar heating unit will vary somewhat depending on the size of your pool and its geographic location. Doing all the work yourself, your expense for materials will range from $600 to about twice that. Most pool owners find that using the sun's heat to warm their pools pays for the installation cost in two or three years. Even if solar energy is used to supplement an existing heater, you'll find that sunlight does most of the work and your fuel bill is no more than about 10% of what you've suffered before.

Heating Your Water Supply

Savings in heating water for your home can be just as dramatic and effective. In Chapter 5 you'll see how you can take your choice of numerous solar systems, available right now with off-the-shelf hardware from manufacturers in many parts of the United States and some foreign countries. One of the most economical kits gives you a solar collector panel and all the additional hardware needed except piping and a few plumbing fittings for about $400. Another package includes two solar collectors and all hardware, including pipes and nipples, for about $700. In both cases, these systems are designed for easy connection to your existing conventional hot water heater. Most solar installations which retain an auxiliary standard heater save the owners at least 70% in fuel costs.

In areas with a great deal of sunshine, such as the entire southern half of the United States, there are already thousands of solar water heaters doing a good job. Owners report fuel bills for heating domestic water are as low as $1 a month. Many installations where there is no electricity or natural gas find solar water heating to be the comfortable and economic answer. To show how important solar water heating is becoming, Florida has had a law in effect since October 1974 requiring that all single family residences must be plumbed for easy installation of solar heaters.

It's possible to heat water in your home by sunlight for well under $1,000 depending on how much individual effort you're willing and able to provide. Take a look at your utility bills. Figure how much you spend each month for heating water. In the average home the cost is at least $15—and rising, how rapidly no one knows. This means that you can pay for a solar water heater in four to seven years, or less.

Heating Your Home

As mentioned before, close to 20% of the total U.S. energy consumption goes for heating homes and other buildings, public and private. One of the announced objectives of the Energy Research and Development Agency is to encourage as much use of solar energy for space heating as possible, in order to conserve fossil fuels. It's possible that in ten years, with full cooperation of government, industry and the public, we'll be able to save half the fuel that goes up the flue.

There's an exciting acceleration in the use of solar energy for heating homes. Two years ago there were only about thirty "solar homes" in this country. Many of them, like the four Massachusetts Institute of Technology (M.I.T.) solar houses in Massachusetts, were built and operated primarily as laboratories to experiment with various techniques for building solar systems.

By March 1975, the number of buildings in the United States heated by solar energy had reached one hundred. And this total was beginning to grow rapidly, despite the deep economic slump in the housing industry. Architects and builders announced plans for solar homes and began constructing them. A developer in Arizona said his new tract would include solar heating and cooling in every home.

Now it appears that by the end of 1975 there may be as many as four thousand homes equipped for solar space heating. Public interest has been fanned from a spark to a solar flare by stories in newspapers and national magazines and on television and radio programs. Estimates of government funding for solar projects ranging from solar homes to huge, non-polluting solar power plants, have gone up from less than $100 million to more than double than amount in the fiscal year 1976. It is likely that this budget will go still higher as more people realize the immediate and future gains easily available by using existing techniques to benefit from sunlight.

Two Solar Heating Methods

There are two basic methods for using solar energy to heat your home, one using liquid (usually water) as the heat transfer medium; the other using air. Both are capable of carrying from 50% to 90% of the heating load, depending on where you live.

4

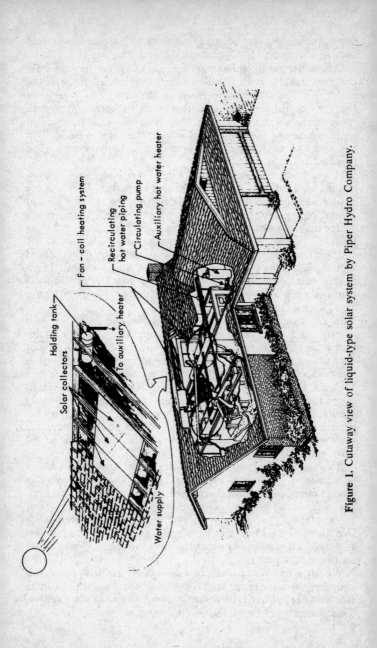

Figure 1. Cutaway view of liquid-type solar system by Piper Hydro Company.

Solar collectors

Holding tank

To auxiliary heater

Water supply

Fan-coil heating system

Recirculating hot water piping

Circulating pump

Auxiliary hot water heater

Liquid-type solar systems

One technique for home heating is to use water in your solar collector panels. Then this hot water is circulated through the building, giving off heat and providing hot water. For space heating, the fan-coil approach also works well. A thermostat in each room automatically turns on a fan which blows air over coiled pipes containing the sun-heated water. The same pipes deliver hot water to your bathrooms, kitchen and laundry area. (See Figure 1.)

If your home is in an area where there is freezing weather, you can use an antifreeze mixture in your solar panels. Then the hot antifreeze circulates through a heat exchanger—a series of coils of heated piping connected to the solar panels and placed either within the hot water storage tank or in a small insulated metal box outside the tank. Water from the storage tank flows around the coils (or into the box), is warmed by this contact, and then stored in the tank. By the end of the day, you have a big tank full of sun-heated water. After a few days of sunshine, your storage tank will continue to heat your house and provide hot water through a cloudy spell lasting several days.

In any case, whether using water or antifreeze as your heat transfer medium, it has been proven most efficient to use a liquid-type solar system to supplement a conventional heating system of the hydronic (liquid) type. Solar energy in a well designed installation should cut your fuel bills by 60% to 70%.

You'll find information in later chapters as to how to install such a liquid-type heating system. Unless you've had considerable plumbing experience, you'll probably want the services of an experienced plumber at times. However, there's no reason why you can't select and buy the necessary materials, and do most of the work yourself.

Because installing a liquid-type solar system does require a fairly substantial amount of piping and other plumbing hardware, storage tanks, a heat exchanger, pumps and a differential thermostat, this type of system is most suitable for new buildings. So if you're planning to build a new home, be sure to give careful consideration to a solar liquid-type heating system. This approach also makes it possible to cool your home in the summer, as described in Chapter 6.

While the hydronic technique is most easily applied in new construction, there are many kinds of existing buildings which can be efficiently retrofitted with a liquid-transfer solar heating system. These criteria are explained in Chapter 4.

Estimates of Cost for Solar Liquid-Type Heating Systems

Costs of a solar heating system of the liquid type vary widely depending on such factors as:

1. The size of your house and its architectural design; whether it's one or more stories.
2. Geographic location of building; number of days of annual sunshine.
3. What percentage of the heating load is to be furnished by solar energy.
4. If retrofitting an existing building, problems in placing piping, solar collectors, storage tank and other hardware.
5. Possible tax advantage in some cases because solar energy is used; increased taxes in other areas because of added value to the building.

Estimates of what it will cost you to install a solar hydronic heating system can only be made after you've selected the hardware you want, figured how much expert labor is needed beyond your own, and considered the factors listed above. For new construction, estimates of the added cost of a solar system range from under $2,000 to as much as $8,000 for a large residence.

Typically, if you're building a house costing $40,000, you might expect your solar system to add $3,500 above the price of a conventional heating system. Your solar system will heat your house and your domestic hot water. It will add about $750 to your down payment on the house, but it should save you at least $45 a month in operating expenses at present fuel costs.

It's only logical to assume that fuel costs will continue to rise. Studies made by financial experts indicate that if you save $45 a month in fuel bills, and the annual rise in fuel costs averages 5%, and if the interest rate you pay on your solar installation is 10%, you can afford to invest about $4,300 in your solar heating equipment. At the end of ten years, it will be paid for. Your house will be worth considerably more because, except for equipment maintenance, and a small electric bill for running two pumps (part time) and a thermostat amounting to less than $1.50 a month, your major source of heat energy is free.

Air-type solar heaters

The other principal type of solar heating system uses air as the transfer medium. Detailed descriptions and drawings of this technique are included in Chapter 4. Using solar collector panels with air blown through them is an efficient way to provide heat in any building having a forced air heating system. You store solar heat energy by passing the hot air through a bin filled with small round stones the size of golf balls or smaller.

There are many advantages to applying an air system if you already have a house equipped with a forced-air heating system. As you'll see, it's now possible to install an A-frame shed in a suitable place in your yard. An example appears in Figure 2. Such a unit, designed by International Solarthermics Corporation, includes solar collector panels on one side, with 35% greater efficiency because of a reflector shining more sunlight on the panel surfaces. Inside the A-frame structure, which is the size of a tool shed, is a fan to blow the hot air through a sizeable rock bin, and another fan to blow heated air from the bin into a duct connected with your home's ductwork.

Figure 2. A-frame air-type solar heater designed by International Solarthermics.

A system like this can be added to almost any existing home if you have forced-air heat and space to put the A-frame shed so that the side having the solar collector area faces south and is not shaded from the sun during the seven months of the year when you need heating.

Costs of a system like this range from about $2,300 to over $4,500 depending on factors previously mentioned—the size of your house, geographic site, and how much labor you provide yourself. So the same kind of economics cited for a liquid system apply. Your heating bills are cut by 70% and your solar heating system should pay for itself in six to ten years.

Life of Solar Systems

With the same kind of maintenance you normally give to your home or apartment building, there's no reason your solar heating system should not be good for the life of the structure. An air-type solar system continues to heat a house retrofitted in 1944, in Boulder, Colorado (see Figure 16). Most manufacturers of solar collector panels and other associated hardware are now willing to provide long-term warranties, providing you take adequate care of the equipment. Suggestions as to how to do this are included in later chapters.

TERMS USED IN THE WORLD OF SOLAR ENERGY

It is helpful to know the meaning of specialized words used in the relatively new and growing business of solar energy.

Insolation is the amount of solar energy striking a surface of the earth. For instance, you might compare the average daily insolation on a wall facing true south versus insolation on a wall facing true north. The northern wall gets only about 40% of the solar energy reaching the southern wall in the northern hemisphere, in the latitudes between 30° and 45° N.

Carrythrough refers to the number of days without sunshine during which the heat storage system in a building heated by solar energy can provide adequate space and/or water heating.

Retrofitting is the process of installing a solar heating system in an existing building, usually to take over the major part of the heating load from a conventionally fueled system.

Downpoint is the temperature at which useful heat energy can no longer be taken from storage. In an air-type solar system, useful heat can be taken from the rock storage bin at temperatures as low as 75° F. In most liquid-type solar systems, the downpoint is considerably higher—at least 90° and preferably about 110°.

Solar collector panel is a sandwich-like structure with a black

9

surface to absorb solar heat. It may be used to heat either liquid, usually water or antifreeze, or air as the transfer medium. Details of various solar collectors, and how to build some of them, are provided in Chapter 3.

Thermal lag represents the time it takes to collect enough heat so that your storage unit climbs beyond its downpoint temperature after a period during which your system has collected no solar heat.

A complete glossary of all technical and scientific terms used in this book can be found at the end of the book.

2 / Choosing the Best Solar System for Your Needs

Even when there's snow on the ground and the roof, solar energy can provide most of the heating requirements for buildings as large as stores and schools (see Figure 38). But in order for a solar system to work at maximum efficiency, several factors must first be considered. As mentioned earlier, the size of your house or building, its geographic location, its existing heating system, and the amount of time and money you can afford to spend on a new system, should affect your choice of solar heater.

There are many solar heaters now on the market, and the number of designs you can build, buy, or combine, is quite impressive. This chapter will help you determine the best solar heating design for your home, based on the above criteria. It will consider the pros and cons of air-type heaters and liquid-type heaters; various types of collector panels and how to position them; and some wide-ranging designs of solar heating systems now in use.

A GROWING ENTERPRISE OFFERS VARIETY

For many years the few scientists in various parts of the world who pioneered methods to harness solar energy were considered to be harmless eccentrics. They struggled along with a few thousand dollars, usually given by big foundations like Ford and Rockefeller. Their laboratories were hidden away on the campuses of various universities. The general attitude of the public was: "Go play with your solar collectors and reflectors. Make your magic but don't bother us practical people."

Now the solar show has moved into the big tent. As you'll see in Appendix I, there are many companies building hardware and supplying instructions for solar heating and cooling systems. Another large group is working to develop solar cells, with substantial financial backing from major oil companies as well as from leading electronics manufacturers.

11

Congress has recognized the need to accelerate solar development. Under the supervision first of the National Science Foundation, and now of the Energy Research and Development Agency (with help from the Department of Housing and Urban Development and the National Aeronautics and Space Administration), some hundreds of millions of dollars are being provided for solar energy research.

Practically, this means that an architect or builder who wants to construct a new school, hospital, apartment complex or shopping center, may be able to get financial aid in installing a solar heating and air-conditioning system. Certainly, they can get a lot of technical help from experts in both government and industry.

For the individual, particularly if you're a do-it-yourself type, this growing public and private investment means that you'll be able to build a more efficient solar heater and cooler for less money.

COMPARING LIQUID AND AIR-TYPE SOLAR SYSTEMS

Advantages of Liquid-Type Solar Heaters

In the previous chapter, you found a brief description of the two major types of solar heating systems, one using liquid as the heat transfer medium, the other using air. It's worth considering some of the advantages of each kind of system.

Here are reasons for selecting a liquid-type system:

First, it permits you to design a more versatile system. You can use heated liquid for space heating, for warming domestic water as well as your swimming pool, and for cooling your home.

Liquid is a better heat transfer medium than air. You can get far more heat from an exchanger where the solar-heated liquid warms water than from an air-to-water heat exchanger. This means that a liquid heat exchanger is far smaller than an air-to-water unit, and correspondingly less expensive.

Pipes used for carrying liquid from your solar panels to your storage tank are much smaller than the ducts used for air-type collectors. Hence it's easier to make this more compact layout of piping within the limits of your building design.

If your building has been designed for space and domestic water heating with a hydronic (liquid) system (see the description of Piper Hydro in Chapter 4), then it may be relatively inexpensive to add solar collector panels of the liquid type. Your cost could be as low as $2,000 for a solar system delivering 80% of your space heating and domestic hot water.

You have a far wider selection among makes of liquid-type solar collectors. Manufacturers of this type outnumber firms producing air-type systems by a margin of about ten to one.

There are far more liquid solar systems installed than those using air flow. Thousands of hydronic systems in the United States and an estimated 300,000 in other countries are used for heating water in homes, cabins, huts and farm buildings, as well as many thousands for heating swimming pools.

Advantages of Air-Type Solar Heaters

There are some strong arguments for considering an air-flow solar system, particularly if you want to use solar energy for heating an existing building which includes a forced-air system:

System compatibility is quite a benefit since you can connect the ductwork from an air solar heater directly to the ducts, blower and thermostats now installed in your home. As long as the ductwork is well insulated, the entire system will be quite efficient.

System installation and maintenance may be somewhat less costly than with many liquid systems. With water as the transfer medium, you must be very careful to build and maintain a system which doesn't leak. Corrosion can be a problem unless you choose materials for handling liquids that won't corrode. In cold climates, freezing is a hazard unless you fill your collector system with antifreeze. If you use antifreeze and a corrosion inhibitor in the collector water, you must include a liquid-to-liquid heat exchanger to transfer solar heat from the antifreeze solution—usually 50% ethylene glycol and 50% water—to your hot water storage tank.

There can be some gain in collector efficiency if you use air. Most hydronic systems are not operated below a downpoint of about 100° F, the lowest temperature at which you can extract useful heat energy from the system. Air solar systems with rock storage provide useful heat energy from the system. Air solar systems with rock storage provide useful heat at temperatures as low as 75° F. (On the other hand, there are new types of liquid solar collectors such as those designed by Corning and Revere, Sol-R-Tech (Garden Way) and Sunworks, plus a do-it-yourself design like Edmondson's SolarSan, that produce superheated water or antifreeze mixture at temperatures up to 300° F. With superheated liquid, your solar system can warm a large water storage tank quite readily, and assure carrythrough of solar heating for several days of bad weather. This arrangement, however, requires a leak-tight, closed-cycle liquid system, as described in Chapter 4).

There is less thermal lag in an air system. This lag represents the time it takes to collect enough heat so that your storage unit climbs beyond its downpoint temperature after a period during which your system has collected no solar heat. Think of it this way. It takes far less time using hot air to bring a rock bin up to working temperature than the time it takes for hot liquid, usually working through a heat exchanger, to bring the water in a big storage tank up above its downpoint after a prolonged non-collecting period. (Again, this argument becomes less powerful as more efficient liquid collectors are used, raising the water temperatures in insulated storage tanks beyond 200° F. In such cases, the storage water can provide useful space heating as well as hot water for several sunless days.)

There is a gain in terms of heat storage when you use a rock bin due to temperature stratification. That is, useful heat can be taken from a stone bin when up to 90% of the rocks are colder than the downpoint, for the top layer of rocks will still deliver useful heat when you blow air through the bin.

Guidepoints in Your Selection

With cogent arguments for both types of solar heating systems, the debate as to their comparative merits is certain to continue for many years. Some manufacturers, such as Sunworks, offer both types of solar collectors. Although Everett Barber, president of the company, has built a solar home near his factory in Connecticut using the liquid-type collectors for heating the house and its water supply, he is also building air-flow panels. These air-type modules, illustrated later in this chapter and in Chapter 3, are lighter and cost somewhat less than most liquid types.

Another anomaly is that the most recent of six solar homes designed by Dr. Harry E. Thomason, a renowned solar pioneer, utilizes liquid-type solar collectors and a large insulated water tank surrounded by a rock bin. When the rooms in this house need heat, a blower drives the cool air down through the basement rock bin. Air rising through the bin is heated and blown back through the house. Hot water for this Thomason house comes from the big water tank surrounded by stones. This installation is described in detail later in this chapter along with a sketch of the heat storage system (see Figure 8). Thus Thomason has combined several of the best features of both liquid and air solar heating.

14

Here are suggestions from an unbiased student of both liquid and air types of solar heating systems. You will understand the reasoning far better after you've read the next few chapters:

If you're interested only in heating domestic hot water, or warming your swimming pool, liquid-type solar collectors and associated plumbing are the logical answer. This applies whether you're heating water for a suburban home, apartment building, or any structure with city utilities, or for your remote cabin far from electric or gas lines.

Should you want to add solar heating to a home or apartment house having piped distribution of hot water for hydronic or fan-coil space heating as well as domestic hot water, then it makes sense to install liquid-type solar collectors.

When you're planning to build a new home, apartment house, store, shopping center, office building or factory, our first recommendation is to consult with an architect having experience in designing solar-heated structures. You can get the names of qualified architects from the American Institute of Architects (AIA) headquarters in Washington, D.C. In the bibliography at the back of this book is a current list of publications on solar buildings available from the AIA.

Once you've studied solar energy systems, you'll be able to contribute many good ideas toward the design of a new building. If you want a combination of solar heating and cooling, plus production of domestic hot water, you'll probably lean toward a liquid-type system using solar panels which produce superheated liquid. In the next chapter are details covering such new and efficient liquid-type collectors.

If you want to apply solar heating to an existing building having a forced-air system, the most efficient solution is to use air-type collectors and a rock bin for storage. Some of these systems are described in Chapter 4.

If you prefer to use a forced-air system in a new building, you may want to design a hybrid solar system similar to Thomason's. This will enable you to have ample domestic hot water as well as a heated rock bin to provide hot air for the rooms in your house. Although Thomason uses his rock bin for storing cool night air in the summer and thus enjoys a built-in air conditioner, for a larger building you could use the superheated liquid from your collectors to operate an absorption cooler, as described in Chapter 6.

By the time you've become expert on solar systems, all kinds of ideas on using both liquid and air techniques will occur to you. Don't think that you must be an erudite scientist to generate useful

15

new ways to use solar energy. Some of the best concepts have come from relatively untrained experimenters with plenty of curiosity, a knack with materials and tools, and a lot of persistence.

BASIC DESIGNS FOR SOLAR SYSTEMS

Liquid-Type System

In Figure 3 are schematic diagrams created by Revere Copper and Brass to show the principal components of a liquid-type solar system.

Looking at the upper drawing labeled "Heating," you see that the heat transfer liquid in the coils of the large storage tank (usually an antifreeze-water mixture containing a corrosion inhibitor) is pumped to the lower end of the solar collector panel array. As the liquid is heated by the sun's rays, it rises to the upper end of each panel. It flows through a header pipe connected to the outlets of the panels and back into the coils of the heat exchanger in the storage tank. Thus, the solar-heated coils warm the fresh water in the storage tank.

This heated water is then passed from the heavily insulated storage tank through a smaller, conventional hot water tank (marked "auxiliary heater"). This smaller tank, usually containing from 30 to 50 gallons of water, is heated, when necessary, by an auxiliary energy source—gas, fuel oil, or electricity.

From the auxiliary tank, the heated water, as shown in the diagram, can be used in either of two ways for space heating. If your house has a hydronic system with hot water pipes in the floor, baseboard, walls, or ceiling for radiant heating, then the hot water circulates through these pipes under the control of your thermostats. If you prefer a warm air system, then the lower right corner diagrams a fan-coil system. Air is blown across coils of pipe carrying solar-heated water.

Only if there is a prolonged spell of cloudy weather does the auxiliary heater using conventional fuel become the principal source of hot water. However, there is unanimity among solar heating experts on this point: it is economically sound to design your solar system to carry from 60% to 90% of the heating load and to provide an auxiliary heat source using conventional fuel for use during prolonged periods of cloudy weather.

The percentage of heating your solar system will provide will vary with the location of your home. In New Hampshire it's uneconomical in general to design a solar system providing more than 60% of the heating. With a house in Arizona, Southern Cali-

16

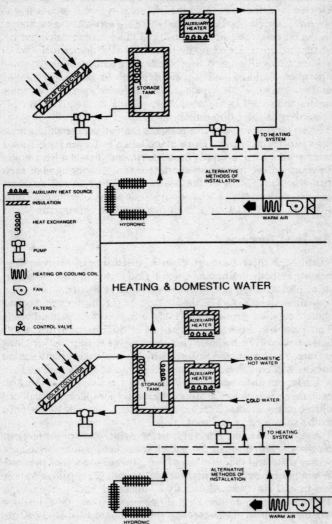

Figure 3. Designs for liquid-type solar heaters.

fornia, or Florida you might have to use the auxiliary heater only 10% of the time or less.

Looking at the lower diagram in Figure 3, labeled "Heating & Domestic Water," you'll see that the solar system is identical with the one just described except that there is a second heat exchanger in the big hot water storage tank, and this exchanger delivers hot water through a second auxiliary heater to the domestic hot water system. Such a design is only needed for a large house or small apartment building having heavy demands for both space heating and hot water. For a typical six-room house, a single auxiliary heater connected to the main storage tank is adequate for both space heating and domestic hot water.

For a large apartment complex, the fan-coil version of this heating system is preferable. Hot water is taken by feeder pipes from a main piping loop to supply each apartment's needs. There would be room on the roof of an apartment complex for enough solar panels to carry most of the heating load, but several auxiliary heaters at regularly spaced intervals would also be needed in a structure containing more than one hundred apartments.

Air-Type Systems

Although there's a more detailed discussion of air-type solar heating systems in Chapter 4, you'll find it easy to understand the principles of a typical system by looking at Figure 4.

In these drawings, (1) is the air-type solar collector panel on the slanted roof. It consists of a flat steel absorber plate painted black. This plate has a cover of two sheets of double-strength window glass, spaced ¾ inch apart. This cover serves to trap the very long infrared solar radiation absorbed by the black steel plate so that this heat energy is not radiated back into the atmosphere.

Under the absorber plate, which also acts as an air-to-air heat exchanger, is ½ inch of air space. Below the air space is a thick pad of insulation, at least 2 inches of fiberglass or rock wool fastened to the sheathing of the roof.

As indicated in the drawings, air from the rooms of the house is blown upward, through the ½-inch space between the heated absorber plate and the insulating layer, toward the top of the roof. When the house is cold and the storage bin (2) is cold, this now sun-heated air is carried directly by ductwork back to the rooms. This action is controlled by a differential thermostat (3), which will be explained in Chapter 4. If the rooms are adequately warm, however, the hot air is then directed to the heat storage bin (2). This dry storage unit is full of dry pebbles or small rocks. These

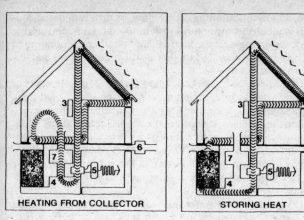

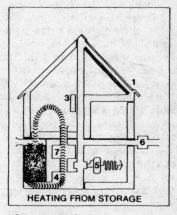

LEGEND

1. SOLAR COLLECTOR
2. DRY STORAGE UNIT HOT AND COLD
3. CONTROL UNIT
4. AIR HANDLING MODULE
5. HOT WATER UNIT
6. DAY-NIGHT EXCHANGE COOLER
7. AUXILIARY UNIT

Figure 4. Design for air-type solar heater.

pebbles have been washed clean before they're put in the bin. Otherwise, you'd have dust from the stones blowing all through your forced-air system. On days when the sun isn't shining to heat the collectors, and the rooms are cold, heat can then be drawn from this storage bin. The average solar-heated storage bin can carry the heating needs of a six-room house for two to three days in cloudy weather.

All heat storage and distribution is controlled by the unit labeled (3), an air-conditioning control box, or differential thermostat. This is an electronic control system to maintain comfortable temperatures throughout the house in any weather, responding to temperature inputs provided by thermistor sensors located at appropriate points.

In the basement space marked (4) are placed all the elements of the air-handling module, including connecting ducts, air filters, blowers, and automatic dampers, which cut off the air flow from a room when the thermostat registers the desired temperature. Usually this module is next to the pebble storage bin. In a house without a basement, the same effect can be achieved by putting items (2), (4), (5) and (7) in a utility room. Or the entire system can be outside the house, as with the International Solarthermics A-frame shed shown in Figure 2.

Item (5) is a solar hot water unit consisting of an air-to-water heat exchanger and a preheat storage tank, which is then connected to a conventional hot water heater (7). As you've seen in the liquid-type heating system, this auxiliary heater consumes fuel or electricity only when solar energy has not heated the water to the desired preset temperature, say 140° F.

Indicated by (6) is a day-night exchange cooler of a type described in Chapter 6. This space cooling system is usually inexpensive and quite effective if the system is properly designed.

Because no solar heating system—except for some in areas with useful isolation throughout the year—can be built economically to provide 100% of your home's heat and hot water, item (7), the auxiliary heater, is needed. It is quite possible to design a solar heating system, air- or liquid-type, to carry 100% of the load. But it would cost you too much because it's overdesigned. In many regions you'd have to make the heat storage facility gigantic to handle a carrythrough of ten to fourteen days of sunless weather.

SOLAR COLLECTOR PANELS

Air-Type Panels

In the description of item (1) in the air-type solar heating system, the various parts of an air-type heat collector were introduced. Reviewing this structure, going from the outer surface to the layer against the roof, this sandwich consists of: two sheets of glass or reinforced clear plastic, separated by ¾-inch air space; another ¾-inch air space between the lower pane and the metal solar collector plate, painted black to absorb the sun's heat; then another ½-inch air space below the metal collector and above the thick pad of insulation firmly attached to the roof.

Air heated by the absorber plate is blown through the ½-inch duct between this black plate and the insulated roof.

Liquid-Type Panels

A typical liquid-type solar collector panel is fairly similar in construction, with some noteworthy differences. A typical sandwich consists of: one or two sheets of glass or transparent clear plastic (if two panes are used, then there is about a ¾-inch air space between them); another air space of ¾ inch above a metal absorber plate painted black. This plate contains, either attached to it or built-in, black metal tubing to carry the liquid heated by the sun. There are usually header pipes at the top and bottom of each collector plate, the lower one to bring in cool fluid, the upper one to carry away sun-heated liquid.

Directly below this liquid-carrying absorber plate is a heavy pad of insulation such as fiberglass board. All these layers are sealed together in an open wood or metal box, or clamped at the edges by a rigid metal "T".

There are many variations in air and liquid solar collector panels. These varieties are described and, in some cases, illustrated in Chapter 3.

How to Mount Your Solar Collector Panels

There are many possible locations where you can mount solar collectors. Several of these alternatives are shown in Figure 5.

Certainly the most frequently used location is (A) for two reasons. One is exposure to sunlight without obstructions. The other is that the collector panel, if suitably designed, can be flush-mounted and provide part of the insulated surface of the roof.

In (B) the collector is away from the roof edge. This may appeal to you if you object to seeing the panels from the ground. Or you may have a flat roof and need to mount the collectors on a tilted rack. In any case you can improve collector efficiency by putting a flat sheet of polished aluminum foil mounted on plywood along the foot of the collector. This acts as a reflector of sunlight and can improve collector performance by 25% or more.

A similar location is the sawtooth arrangement in (C). This has been used successfully on the roofs of large single-story buildings. Some schools, for example, have been equipped with solar heat-

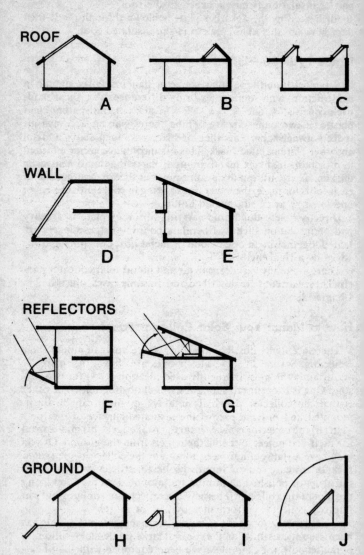

Figure 5. Alternative locations for solar flat plate collectors.

ing and hot water systems using this technique for mounting panels. If you use this approach, be sure the rear collectors are not shaded by the front array, particularly in the winter in northern latitudes, when the sun is low on the southern horizon.

For installations above 42° North (or South) latitude, it's quite feasible to mount your collectors on a wall as in (D) and (E). This is exceptionally effective in areas where there is snow cover for long periods in the winter, because the snow does not cling to the wall, and the snow on the ground serves as a useful reflector. Also the overhang in (E) can be a useful sunshade in midsummer and prevent overheating of collectors at that time.

Two of the many possible approaches to using reflectors are shown in (F) and (G). These locations improve collector efficiency. Arrangement (G) has been tried by Barber and Watson in Connecticut (see Bibliography). They report that, averaged over clear and cloudy days, twice as much insolation reached the panel in arrangement (G) than in arrangements without reflectors.

(H) is one style of ground-mounted collector, while (I) shows a cradle mounting which can be adjusted either manually or automatically so that the inclination of the panel array will change with the season. Note that this latter mounting requires flexible couplings of either piping (for liquid-type systems) or ductwork (for air-type) between the collectors and the house.

(J) shows a glass or plastic enclosure for the collector array. This enclosure serves as the first insulating layer. It might be used with a large portable structure, like a portable radar shack or a geophysical prospector's carry-around hut; or where you want to protect your solar collectors against both weather and possible vandalism.

Correct Angle is Important

Various angles of inclination suitable for mounting collector panels in the Northern Hemisphere are shown in Figure 6. As a rule of thumb, your solar collectors should be at an angle from the ground of at least 10° plus your latitude. Thus, in New York, latitude 41° North, your panel array should slope at least 51° from the horizontal. For Chicago, the slope is about the same. For Los Angeles, latitude 34° North, the preferred angle is 44°.

The further north you go, the more important this inclination becomes. Thus a recommended inclination for Boston (42° North) adds 15° to the latitude and becomes 57° from horizontal.

In the Southern Hemisphere, all solar collector panels face north and the angles of inclination are the same.

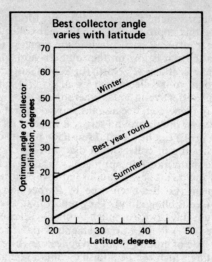

Figure 6. Determining optimum slope for collector panels.

Compass Orientation

The most effective direction in which to aim your solar collector array is about 10° to 15° west of true south. You'll find this best because afternoon temperatures are frequently 20° warmer than morning temperatures, and thus pointing your collectors to take full advantage of afternoon heat is better than aiming due south.

You can readily find the correct orientation by using a compass. However this will point to magnetic north. Therefore you must know the correction in your area to determine where true north is, and of course true south is 180° away from this direction. There are charts for each geographical area which show the variation, either east or west, between magnetic north—the direction which a compass needle says is north—and true north, which is the geographic north from that specific location. If the compass you are using is near magnetic fields of any strength, its accuracy will be further deteriorated by a deviation from the magnetic north. Detailed discussions of this subject are found in many physics texts and in pages 250-268 of the sailor's classic, *Piloting Seamanship and Small Boat Handling* by Charles F. Chapman, 1969-70 edition, published by Motor Boating, New York.

SAMPLE HOMES USING COLLECTOR PANELS

Early Solar Heating Installations

Probably the first buildings heated by solar energy in the United States were cave dwellings in New Mexico and other parts of the Southwest, where Indians used primitive reflectors to beam sunlight indoors.

Certainly one of the first buildings, or group of buildings, to use solar heating techniques similar to present designs is the national monument known as Scotty's Castle, at the northeast end of Death Valley, California. Some two million dollars is the estimated cost of the buildings, which include a main two-and-a-half-story castle designed by Alexander McNeilledge and paid for by Albert M. Johnson, a wealthy Chicago insurance executive. There are two collector arrays, one high above ground, one in a separate building shaped like a greenhouse on a shelf of land above the main ranch house. Collectors consist of copper pipes ¾ inch in diameter, painted black, behind a double thickness of glass. Both arrays face south, and they are slanted about 40° from the horizontal.

This early system, installed in 1929, provided both bathwater and space heating in the castle for Walter Scott (Death Valley Scotty) and his colorful friends.

Solar Houses Developed at M.I.T.

A few years later, thanks to a grant from Dr. Godfrey L. Cabot, several scientists at Massachusetts Institute of Technology built a series of solar houses. Under the direction of Dr. H. C. Hottel and B. B. Woertz, M.I.T. Solar House I was completed in 1939 using 400 square feet of water-type collector panels. The panels contained a black copper sheet on which were copper tubes covered by a triple thickness of glass. The collector array was mounted on a south-facing roof inclined 30° and fed heated water to an underground storage tank containing 17,400 gallons of water. Heat was carried to the two rooms of this "house"—actually used as a laboratory—by air. By the end of the first summer, the thermometer in the big storage tank read 195° F.

Later houses in the M.I.T. series used similar collectors but on roofs with greater slants for more efficiency. Also, thanks to work by Dr. Maria Telkes and Dr. C. D. Engrebretson, two other noted pioneers in solar engineering, a more sophisticated storage system was developed. This technique used bins containing eutectic salts, which work in an unusual way. Eutectic materials have the valuable property of remaining at a constant temperature while absorbing heat energy, by going from solid to liquid, or while giving up heat energy, by going from liquid to solid.

In this case, Glauber's salts, a common chemical, were used, and transformed by the solar heat from a solid to a liquid. This eutectic transformation of Glauber's salt occurs at a temperature easily reached by the antifreeze mixture in solar panels. Then the eutectic bin stores heat until the salt returns to solid crystals, giving up its heat to the surrounding air. If air is blown over the bin, you have a hot air furnace operating on solar energy.

A house in Cambridge, Massachusetts, heated in this way, obtained 90° of its energy for both space heating and hot water from the sun. Even during the four coldest months, 75% to 85% of the total heat needed was provided by liquid-type solar collectors, with heat stored in eutectic bins.

Thomason Solar Houses in Maryland

Six solar homes in Maryland, a few miles from Washington, D.C., have been built by Dr. Harry E. Thomason, aided by his son. These have been so successful that quite a few other builders of solar houses are using some of Thomason's techniques.

The senior Thomason, long a patent attorney for the government, got his original inspiration while standing in a shed with a rusty metal roof during a rainstorm in North Carolina. He noticed that the water running off the corrugated roof was warm, obviously from having absorbed solar energy.

Using this principle, the Thomasons have designed and built a series of houses where solar heating of the rooms and water in winter, and a utilization of solar techniques for summer cooling, have proved highly successful in slashing fuel bills. The first of the Thomason Solar Homes was built in 1959 and its heating system still supplies 85% of the annual requirements for this three-bedroom home.

In fact, during the first three years of operation of this hybrid solar heating system, Thomason's total bill for fuel oil for his auxiliary furnace came to less than $20. By contrast, similar homes in that chilly Maryland climate cost many hundreds of dollars annually for heating.

Trickling Water Systems for Absorption of Heat

All six Thomason houses use trickling water as the medium for absorbing solar radiation. On his first house there are 750 square feet of solar panels on two areas of the roof, both facing south. Slopes of these roof sections are 45° and 60°—resulting in a high degree of solar efficiency since the latitude is about 30°. Along the ridge of the roof is ½-inch copper pipe with tiny holes, $\frac{1}{16}$-inch diameter, spaced 2½ inches apart. This pipe acts to distribute trickles of water down the valleys of a black corrugated aluminum sheet, as shown in Figure 7. A glass sheet is used as the cover for the panels, trapping solar infrared energy.

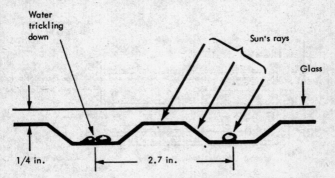

Water trickling down

Sun's rays

Glass

1/4 in.

2.7 in.

Figure 7. Cross-section of solar panels designed by Harry E. Thomason.

Water from the valleys in the aluminum sheets is collected in a gutter and flows to a steel storage tank in the basement with a capacity of 1,600 gallons. One of the innovative features in the Thomason design is that his storage tank, instead of being surrounded with conventional insulating materials, is in a bin filled with 50 tons of stones ranging in size from a golf ball to a tennis ball. (See Figure 8.)

Typically the water temperature in winter in the main storage tank will be 125° to 135° F. This warm water heats the stones surrounding the tank. A blower sends air through the bin of heated stones and into the conventional ductwork of the house. Maintaining a comfortable temperature is accomplished automatically: there are thermistors to measure and compare the temperatures of the roof and of water in the storage tank. While the roof is hotter than the water in the tank, the water continues to circulate. It is driven by a ⅓-horsepower water pump to the perforated distribution pipe of copper, and then pulled back down to the storage tank by gravity. When the roof is cooler than the tank water, the thermistor system turns off the pump, water drains from the roof, and all the water is in storage.

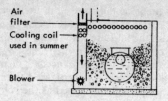

Transverse Section

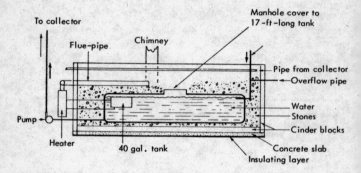

Longitudinal Section

Figure 8. Storage system for solar heat used in the Thomason houses.

Economical Cooling as Well as Heating

Measurement of the temperature in the storage tank is made at the bottom of the tank, where the water is coolest. Because of this automatic system for draining the collector panels when the sun isn't shining, Thomason is able to use filtered water, without anti-freeze, in his system.

Some of the more recent solar homes built by this inventor are larger, but similar in their type of heating and cooling system. The stone bin surrounding the storage tank proves to be useful in summer as well as winter. In the sixth Thomason Solar House, for example, a relatively small standard air conditioner costing $400 is operated in the coolest part of the night, when it is most efficient, to dry and cool the air in the stone bin. (This period, from 10:00 p.m. to 4:00 a.m., is off peak time for electric power loading, a secondary advantage.) During the hot daytime hours, cool air from the stone bin is circulated through the house.

According to the inventor, the total cost of his sixth house—two stories, nine rooms, including an indoor solarium and small pool—is about $65,000 for everything but the land. He estimates that the additional cost for solar equipment to heat the home, preheat domestic hot water, and air-condition the rooms, as compared to conventional equipment, is about $2,000. This sum can be saved in a very few years.

Zomeworks House in Snowmass, Colorado

About 2,000 miles west of the Thomason houses is a novel solar home recently completed in Snowmass, Colorado, about 115 miles southwest of Denver. This is a comfortable house, enjoyed by a young family, where the ingenious solar system was designed by Zomeworks and built by the owner, R. Shore, with the help of a few friends and only $500 for professional labor.

The successful operation of Shore's solar house points to what a dedicated do-it-yourself person can do. We might try to reconstruct his planning and construction, step by step. First, we build the solar collector panels, which now carry the Zomeworks trademark.

How to Build Zomeworks Collector Panels

1. Put a corrugated aluminum roofing sheet on a flat surface. This sheet is 14 feet long and 2½ feet wide, made of standard 0.019-inch-thick aluminum.
2. Above this, put a second sheet of aluminum, identical in every way to the first except that this upper sheet is painted black. The two pieces of standard aluminum roofing nest as shown in Figure 9, providing a space of only $\frac{1}{32}$ inch between them.
3. Over the two nesting aluminum absorber sheets, mount a series of three panes of single-strength glass held away from the metal by ½-inch aluminum spacers—small blocks of aluminum used specifically for separating layers of materials.
4. To provide a second glazing, put three panes of double-strength glass above the first glazing, again using ½-inch spacers as separators. You support both outer and inner sets of panes with aluminum bars having a "T" section.
5. Below the aluminum absorber sheets is 2-inch-thick fiberglass insulation placed in an aluminum frame. You place the assembly of nesting aluminum sheets and double glazing in this box.
6. Mount sixteen of these collector panels on the south roof of your home, making sure they slope at an angle of 10° plus the latitude.

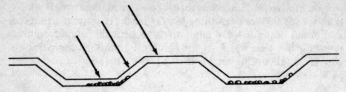

Figure 9. Cross-section of corrugated aluminum sheets in Zomeworks collector panels.

7. Run a header pipe of 1½-inch pipe along the upper edges of the sixteen panels. You have previously attached four ¾-inch pipes for feeders to each collector panel, so that there are now sixty-four feeders (four feeders times sixteen panels) connected to the array of solar collectors. Spacing of these feeders is such that there is one feeder pipe for every other valley in the nesting absorber sheets.

8. At the bottom of the panel array, connect one outlet tube from each collector to a lower header pipe, also 1½ inches in diameter.

9. Now you connect the pipes going to and from this panel array on the roof to a large water storage tank located below ground level. This is a rectangular poured-concrete tank, 6 feet high, 12½ feet long, and 9½ feet wide, with an interior liner of Plas-Chem 5500 butyl rubber sprayed on the inside of the tank to a thickness of 0.030 inches with an airless spray gun. Around this tank, except for the top, provide 5 inches of insulation—3 inches polyurethane foam and 2 inches Dow SM styrofoam. Cover the top of the tank with 2 inches of board insulation, above which, in Shore's house, is the particleboard flooring of the child's bedroom.

   Note that such heavy insulation is needed because Shore's house is at an altitude of 7,200 in the Rockies, 115 miles northwest of Denver. It is cold country in the winter.

10. Except for some additional features invented by Shore, you now have a working system for delivering solar heat to the storage tank. Of course you have connected all the necessary piping and you've provided a ½-horsepower submersible centrifugal pipe. This permits you to pipe water to the upper header pipe. During the sunny part of the day, this water trickles down through the $\frac{1}{32}$-inch space between the black absorber sheet and the sheet just below it. This space is so small that the water is efficiently warmed in this nesting absorber. The warmed water flows at a rate of 10 gallons per

minute, a rate you can achieve by running your pump at about half its rated power.

Connect the collector array to the big storage tank in such a way that heated water from the panels goes directly to the top of the tank. There is no heat exchanger. You feed water from the tank to the collectors in the morning, taking water from the bottom of the tank, which is usually 10° colder than the top of the tank.

The valves in your system are arranged so that you can drain out all the water at night or during the summer, when you can keep them empty and connect a vent pipe to fully dry them out. This summer airing is achieved by disconnecting the pipe carrying solar-heated water from the top header pipe, which runs along the upper edge of most collectors. Now you connect a short L-shaped piece of pipe to the header, in such a way that the open end faces up toward the sky. This open-ended pipe is a vent and serves to exhaust air from the solar collectors, thus helping them to dry. In most colder climates, however, users don't drain closed loop liquid systems because an antifreeze mixture (50% ethylene glycol, 50% water) is used in solar collectors. The antifreeze should contain anticorrosive chemicals and may be used as long as the collectors and associated plumbing are not leaking.

A system like this works satisfactorily, but Shore has added some useful and ingenious new features.

Unique Features of the Zomeworks System

There are three Zomeworks skylids located above the main collector array. These are motor-driven so that each opens at the start of a sunny day, and closes automatically—becoming insulation—at the end of the day or during bad weather. Each Skylid has an area of 16 square feet and gets solar radiation from the planar reflectors faced with 0.018-inch high-reflectance aluminized mylar. Below the reflectors are three additional collector panels like those in the array, except that these collectors are only 8 feet wide and 2 feet high for a total area of 48 square feet. (The main collector array which we "built" has an area of 546 square feet.)

Because of the reflectors, which increase insolation on the small collectors by 25%, these Skylids reach a higher temperature than the bigger panels below and help to keep the water in the big storage tank usefully hot. During the summer, the reflectors open only to a horizontal position and thus act as shades for the small collectors when their additional heat is not needed.

Not only is the insulation of the house excellent, but Shore has also added still another useful feature. His windows facing south are double-glazed and have a combined area of 90 square feet. During sunny, cold days, 60 square feet of this window area is left clear to bring additional solar heat into the house. At the end of wintry days, a blower fills the 3-inch space between the glass panes with about one million tiny white polystyrene beads.

Thus the inventor generates what he calls a beadwall, providing insulation comparable to 3½ inches of fiberglass.

100% solar heating at reasonable cost

With this kind of solar system providing radiant heat to the first floor rooms through ¾-inch polyethylene pipes embedded in concrete, hot water for heating is generally circulated by gravity, since the top of the tank is about 2 feet above floor level of all except the child's room directly above. The latter is heated by energy radiating from the big water tank just below its floor. If necessary, a $1/_{12}$-horsepower centrifugal pump is used to keep water flowing through the floor pipes and back to the bottom of the storage tank.

Since this house was completed in September 1974, Shore's solar heating system has provided all the space heating and hot water needed for it. It is a 100% system. This is one of many exceptions to the rule against 100% solar systems. In Snowmass, Colorado, at a remote 7,200 feet, the cost of auxiliary fuel makes a self-sufficient solar heater highly desirable.

Shore figures that his added cost of materials for the solar system amounted to $3,000. As previously mentioned, he also spent $500 for professional labor. But his solar showplace will pay for itself in three years, and after that, the Shores will have free fuel for the life of their house.

An Air-Type Solar House Built in the Do-It-Yourself Manner

Some fifteen years ago, in Lincoln, Massachusetts, near Boston, another man with handyman talents served as his own architect, builder and solar engineer in constructing an unusual house. Because of his northern latitude, R. W. Gras used his south wall as the solar collector area. If we try to follow his procedure step-by-step, it goes somewhat like this:

1. Start with a cinderblock wall facing south. Gras used an area of about 1,300 square feet because he built a big split-level home with some 3,000 square feet of floor space.

2. At a distance of 5 inches from this block wall, erect a black metallic wall. A series of galvanized steel siding sheets painted black and connected together by suitable framing and bolts would do the job.
3. Erect a glass window—a series of large panes—about 4 inches in front of the absorber wall, to trap infrared radiation.
4. Use the conventional hardware of a forced-air system to blow air from the bottom of the black absorber plates through ductwork, dampers and filters to various rooms in the house.

Because Gras built his home with massive walls and floors, using masonry and hollow concrete beams, with only a few small double-glazed windows, the body of the house has served as an excellent heat storage medium. There has been no need for a rock storage bin or eutectic storage bin.

With the entire windowless south wall open to sunlight and serving as a large air-type solar collector, this house derives an estimated 50% of its heating from solar energy. Auxiliary heat is provided by a gas furnace.

This solar installation, designed and built by the owner, paid for itself long ago. It's a beautiful home for a thrifty New Englander.

3 / How to Build Solar Collector Panels

For most practical applications of solar energy, you must begin with one or many solar collector panels to catch this energy. This chapter will show you how to make several models and will introduce you to the many commercial panels available in kit form or ready-made.

While we associate solar energy with modern science, the idea of using some kind of a blackened flat-plate absorber to collect solar energy is older than our republic. In the second half of the 18th century, a Swiss mountain climber and scientific experimenter—a handy contemporary of Ben Franklin—designed a simple solar collector. Using a wooden box painted black on the inside and covered with two layers of glass, H. B. de Saussure found he had invented a primitive solar cooker.

At various times during the following centuries, inventors in Brazil, South Africa, England, France, Germany, India, Israel, Egypt, Russia, Australia, Japan, and New Zealand, as well as the United States, discovered that they could use a tilted flat-plate collector to heat water above its boiling point. As long as the system was airtight and thoroughly evacuated so that the water was under pressure—and the fittings didn't blow off—water temperatures of 250° F and more could be achieved.

Most of this work was done by experimenters with some degree of scientific training. Then it became desirable to heat water with solar energy in many of the fuel-poor countries like Japan and Israel. Big and small manufacturers began to produce solar water heaters.

And this leads us logically to the current energy crisis and an urgent interest in solar collectors by government agencies and leading political figures, as well as the average, utility-conscious homeowner. So how do you build a solar collector panel?

SIMPLE HOMEMADE DESIGNS THAT WORK

If you remember the brief discussion on liquid- and air-type solar collectors in Chapter 2, you know that there are advantages and disadvantages associated with both types. Also, you're aware that there are a wide variety of liquid types.

The Japanese Solar Heating Tray

To illustrate how easy it is to build a solar panel for heating water, consider a simple design which has been used by Japanese farmers for years. In that fuel-scarce country, many thousands of farms use solar heat to keep buildings warm, heat feed for animals and chickens, and provide warm domestic water.

To construct the Japanese solar collector, build a large wooden tray, using a sheet of plywood 6 feet long by 3 feet wide for the bottom. Sides consist of two-by-sixes. Then you line this box with a thick sheet of black polyvinyl plastic. You insert an inlet tube for cold water near the bottom of the tray, and an outlet tube for hot water near the top. Hot water, being less dense than cold, will rise to the surface.

Your trap for this system is a hinged wooden frame like a picture frame. In this frame are one or two sheets of glass or transparent weather-resistant plastic. If you use glass, you'll probably want to have a double frame over the tray, because a light sheet of glass 6 feet by 3 feet may be too flexible and easy to break. Sealant around the cover is essential.

Many designers consider that a double cover of glass or plastic is more efficient than a single sheet because, with an air space between the two layers, there is less moisture condensing on the underside. Water droplets scatter sunshine and reduce the amount of heat reaching the water in the tray. However, Everett Barber of Sunworks, an experienced designer of modern solar panels and houses, states that a single glass cover does a better job on a collector than two covers.

Inside a typical Japanese solar heating tray is about 45 gallons of water, or a depth of 4 inches. If this tray is mounted on the south side of a roof of a house, shed or barn, with an inclination of at least 30° from the horizontal, then it is a fairly efficient heat collector. Sunshine will heat the water in such a tray on a Japanese farm to a comfortably hot 130° F in summer and about 95° in winter.

An Improved Version of the Japanese Solar Heater Tray

For a still more efficient panel that's easy to build, make your entire tray of galvanized sheet steel. Make it strong enough so that it will hold water when tilted at an angle. Then paint the bottom and sides inside the tray with a selective black paint. A selective paint absorbs more of the infrared wavelengths of sunlight than a non-selective paint. Such a selective blackened surface has an absorptivity of 90% for solar radiation, and an emissivity in the long-ray infrared thermal region of the spectrum of about 10% or less. Hence, a selective absorber improves the efficiency of the collector in general.

After you've made a watertight metal tray and painted the inside, set it in a wooden box. Typical dimensions of the box are 8 feet long by 3 feet wide. The wooden box should be lined with fiberglass, gypsum, rock wool or some similar insulating material. Provide an inlet pipe for cold water at the bottom of the tray, and an outlet for sun-warmed water at the top. Cover the box with a hinged wooden frame containing glass or a suitable plastic such as 4-mil-thick Tedlar.

One way of achieving greater heating efficiency is to mount this boxlike panel on three legs (one in back, two in front). By adjusting the height of the three supporting legs, you can tilt your solar water heater to get the greatest amount of direct sunlight, winter and summer. The object is to have the sun's rays striking the black absorber panel as close to a 90° angle as possible.

Most solar panels in actual service are somewhat complex but utilize the same basic principles as the tilted tray full of water. In the following pages are described some designs for solar panels, many of which you can build entirely or partially yourself, which are functioning satisfactorily in various installations in the United States.

The Thomason Collector Panel

Suppose you want to build a collector panel of the type used by Thomason on his six solar houses in Maryland, mentioned in the previous chapter. This would be the approximate step-by-step procedure:
1. Take a single sheet of standard corrugated aluminum roofing 16 feet long, 4 feet wide and 0.019 inch thick. The grooves in this sheet are 1 inch wide by ½ inch deep and their centers occur every 2.7 inches. Set this sheet on sawhorses and paint the

upper side with selective black paint. A good brand is 3M Nextel Black Velvet over a light gray Nextel primer.

2. Build an aluminum tray, using panels about 4 inches wide for the sides and an aluminum sheet for the bottom. Set the black-painted collector panel inside it.

3. Mount eight glass covers, each 2 feet long and 4 feet wide, in aluminum frames which rest on the black corrugated absorber sheet.

4. Run a piece of copper pipe, ¾ inch in diameter, as a header through the top of your aluminum box. If you're planning to install several solar panels adjacent to each other, make this header pipe long enough to carry water to all the panels; that is, slightly more than 4 feet of header (slightly more than the width of each absorber, for the tray enclosure must be bigger) for each panel. Provide for a vent pipe at the end away from the water inlet to this header. To this header you attach a smaller copper pipe, ⅜-inch diameter, and connect it so that it runs parallel to the header, across the absorber. This 4-foot feeder should be drilled with 1/16 -inch holes centered 2.7 inches apart. This is so that water from the header, flowing through the feeder, will trickle through the holes and down the valleys of the black absorber plate.

5. Under the absorber, between it and the aluminum tray, provide at least 2.5 inches of fiberglass insulation.

6. At the lower end of the tray provide a gutter for the water that trickles down the absorber valleys. With panels side by side, connect these gutters with copper pipe.

Now you have made a solar collector panel similar to the ones used by Thomason on his solar houses. In this design, contrary to most of the liquid-type collectors, the feed water from a storage tank (see Figure 3, Chapter 2) is pumped to the header at the top of the collector panels. From there, the sun-heated water trickles down and feeds into the warm top of the storage tank by gravity or pumping action.

This example of how to build a solar collector is merely illustrative. Dr. Thomason's designs are well protected by patents. If you want to buy plans for his solar homes and a license to build any of them, the information is available for a reasonable fee through Edmund Scientific Company, Barrington, New Jersey.

Individual Variations of Thomason's Collector Panel

It's obvious that designers of solar homes are using trickling-water panels without following Thomason's designs slavishly. An example is Shore's house in Colorado with its many innovations described in Chapter 2.

Another instance is the small house in Bar Harbor, Maine, designed by E. McMullen and owned by R. Davis. A major difference in McMullen's design is the use of a double glazing of 0.040-inch-thick Kalwall Sun-Lite polyester-reinforced fiberglass on the collector panels to trap the solar radiation. These covers are lighter and more resistant to shattering than glass.

A further difference is that there is a backing of 9 inches of fiberglass insulation behind the blackened aluminum sheets. This system, which includes a 2,000-gallon cylindrical steel tank in the utility room, has 540 square feet of solar collectors on a roof facing 10° west of south and slanted 54° from the horizontal. Surrounding the storage tank is a heavily insulated wooden bin containing 40 tons of stones. A tank for water is kept inside the big tank, so that both space heating and hot water for a 1,340 square foot house are furnished by solar energy. The only auxiliary heating is provided by a large wood-burning stove.

The Thermosiphoning Solar Collector

Most liquid-type solar collectors are not of the trickling-water variety just discussed. Instead, the panels contain some kind of pipes, straight or serpentine, attached to the black absorber panel. Or the water may flow through extruded channels built into the absorber plate, as shown in Figure 10.

Still another way to make a panel is quite easy for a home-im-

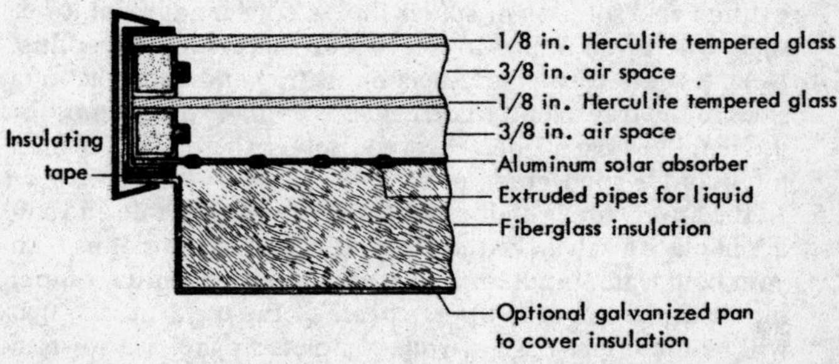

Figure 10. Cross-section of solar collector by PPG Industries.

provement handyman. You build your own channels by using a corrugated sheet of steel as the absorber plate to carry the solar-heated water. This kind of collector, like the many variations having pipes, is called a thermosiphoning unit. It works on the principle that hot water rises. Supply water is introduced at the bottom of the panel and, as it is heated by the sun, it rises to the top of the panel and flows out. In many installations you don't need a pump at all with this kind of water heating system, as long as there's enough water pressure from your supply to bring colder water up to the collector. To achieve this, you put your hot water storage tank higher than the top of your collector array and let the sun's heat force the water into it. This is shown in Figure 49 in Chapter 7 and described in more detail.

Instructions for Building a Liquid-Type Collector

Now let's build a liquid-type solar collector which will provide enough hot water for a small family while the sun is shining:

1. Get a sheet of galvanized steel corrugated roofing. A typical size is 8 feet long and 2 feet 3 inches wide (96 inches by 27 inches). (If your preference is for a lighter panel, use corrugated aluminum roofing sheet in any standard size. Most of the procedures are similar.) Square off the ends of this sheet.

2. Buy a slightly larger flat sheet of 26-gauge galvanized steel. You may want to bend this sheet yourself, using a hardwood block and mallet, to make a tray to contain the absorber panel of corrugated steel. Otherwise buy both the flat and corrugated sheets from a sheet metal company. Have the company bend the ends and edges of the larger flat sheet so that it makes a tray large enough to contain the corrugated absorber. The depth of this tray is 2 inches. See Figure 11.

3. Clean the edges of the corrugated sheet with a wire brush and a dilute acid solution or solder flux. Also clean the flat lower sheet where you intend to solder it to the corrugated roofing.

4. Next, put the corrugated sheet on its tray and bend the long sides of the tray over the corrugations to make tight seams for the length of each side. Then you solder these seams with a torch and a good grade of solder.

5. At the lower corner of one end of the tray, drill a hole and solder in place a galvanized steel nipple threaded for mating with your cold water input line. Next you provide a similar output nipple at the opposite upper corner of the tray, the one that will be the upper edge of your completed panel as shown in

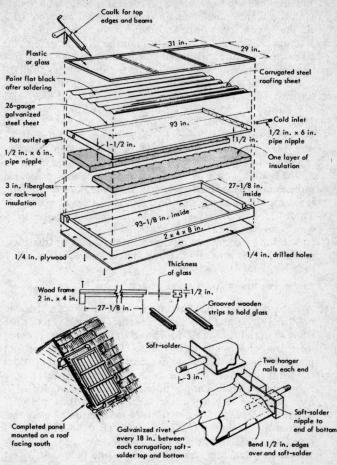

Figure 11. Materials for making your own liquid-type solar collector panels.

Figure 11. Now solder the end of the corrugated sheet to the end of the tray, top and bottom.

6. Now you need a riveting gun. Travel along the valleys of the corrugated sheet, riveting it to the bottom plate of the tray at 2-foot intervals. Then seal each sheet at the rivets with solder.

7. You are ready to test your panel for leaks. Fill it with water. Unless you are a phenomenal worker, there will be a few. Circle them with red crayon and then fix them with solder.

8. You now have a collector but it's not the right color. So you paint the top of the corrugated sheet with a selective black finish. Paints recommended by William B. Edmondson, publisher of *Solar Energy Digest* and inventor of the SolarSan collector panel, are the 3M Nextel Primer 911-T4 Light Gray and 3M Nextel Black Velvet Coating 101-C10 for spraying or 110-C10 for use with a brush or roller.

9. The rest of the job is relatively easy—you're more than halfway there. As shown in Figure 11, build a wooden frame to contain your collector panel. You might use ¼-inch plywood for the bottom, and 2-by 6-inch redwood sides and ends. Drill three evenly spaced holes ⅜ inch in diameter along each of the long sides of your wooden box, and two holes at each end. These holes are necessary to vent any water or condensation which may escape from the metal panel.

10. Now fill the wood box with a pad 3 inches thick of fiberglass or some similar insulation. This will help to retain as much of the solar heat as possible.

Plastic versus glass panel covers

11. Around the top of the box attach grooved wooden strips to hold a layer of glass or suitable plastic. If using glass as the cover material for your 8-foot panel, it's desirable to provide three windows, as shown in Figure 11. That's because 8-foot panels of tempered glass are expensive and too flexible under conditions of severe wind or snow. You can provide a longer panel of the translucent plastic sheets: two windows for an 8-foot panel are safe if you use the 0.040-inch-thick plastic.

Suitable durable plastic sheets with good transmission characteristics for sunlight include Glasteel's clear fiberglass laminated with Tedlar, and Kalwall's polyester reinforced fiberglass called Sun-Lite Regular and Sun-Lite Premium. These plastic covers for solar panels have some advantages over glass. Their solar properites are as good as glass, and they are lighter in weight—typically 0.3 pound per square foot for plastic versus 1.6 pounds for double-strength glass ⅛ inch thick. Also, the plastic collector covers have superior impact- and shatter-resistance yet they can easily be cut with hand tools. Best of all, there is a significant cost saving in using plastic sheeting.

If you wish, instead of a single layer of plastic or glass, use two layers with an air space of about ¾ inch between layers. As to your choice of using one or two layers of cover material, you will gain somewhat in insulation and therefore in heat retention with two layers, according to most experts on solar panels. At the same time, you lose in transmission of sunlight to the black absorbing surface. A single layer will transmit about 90% of the insolation, or incident solar energy. A typical two-layer cover will reduce transmission to approximately 80% while trapping more reflected infrared radiation and therefore heat.

Our recommendation is to use a single layer of suitable plastic or glass because most commercial manufacturers of solar panels have gone to one layer. Obviously it's easier to construct and lighter in weight as well as offering a small saving in material costs. As solar manufacturer Everett Barber states: "The annual thermal efficiency of a single-glass-cover collector with selective black is superior even to one with two glass covers and comparable selective black at temperature differences between absorber and ambient [surrounding air] below about 150°."

12. Whether you select plastic or glass as the cover material for your panel, it's important to seal around the wood frames previously described so that the cover is watertight. A good material for sealing cover panes is silicone caulking compound. If the windows of your solar panels leak when it rains or under any conditions of high humidity, the water collecting on the underside of the cover sheet will greatly reduce solar transmittance. Immediately your 90% transmittance will drop and far less sunlight will reach your black heat absorber.

The Efficient and Easy-to-Build SolarSan Model

You can get better heating efficiency from a liquid solar collector designed by William B. Edmondson, publisher of *Solar Energy Digest*. Another major advantage of this panel, according to its designer, is that you'll find it easy to assemble with simple hand tools. Furthermore, it is considerably lighter than the collector using galvanized steel sheets.

Since a patent application has been filed on this design, our description of Edmondson's collector must be brief.

He calls his design by the trade name SolarSan. As a cover material, he uses not glass but 4-mil Tedlar, the clear polyvinylfluoride plastic sheeting made by Dupont which has good durability in all kinds of weather.

Under this plastic cover, Edmondson's solar sandwich has either a 1-inch layer of fiberglass filter material or an air space. The filter pad transmits a substantial amount of sunlight and provides a spring-like support for the Tedlar film, as well as trapping some of the heat reflected by the collector plate. The inventor states that even when the absorber plate reaches 300° F, the top plastic film feels cool to the hand.

A novel feature of this design is the use of a 3-millimeter thickness of soft, highly reflective aluminum foil as the plate which absorbs or collects the sun's heat. Fastened to this foil is a serpentine of ¼-inch copper tubing. Staples are placed at 1-foot intervals and driven into a fiberglass board 1 inch thick, which is used as backing for the foil. This presses the copper tubing into the foil and makes a tight contact between the two heat-absorbing materials.

Both foil and tubing are coated with a primer, then black graphite paint, and finally a selective black such as 3M Black Velvet, which is an excellent absorber of solar energy.

In a typical installation, this section of the SolarSan panel forms the upper part of a shallow wooden box containing a 4-inch insulating layer of foil-faced fiberglass. A completed panel 8 feet long, 2 feet wide, and 5¾ inches deep weighs 52 pounds, or 3.25 pounds per square foot. The collector area is almost 15 square feet.

According to Edmondson, this panel will deliver water at temperatures up to 250° F if mounted perpendicular to sunlight striking it at an ambient of 80°. Although the inventor has sold some handmade SolarSan panels for $160 each, he would prefer to provide a manual entitled *Solar Water Heaters and Their Application*. This 54-page book provides detailed diagrams for making SolarSan panels and some information as to how they can be used in systems for heating homes and domestic hot water. Price of the book is $26.50. Using it, and not counting your own labor, you might build a SolarSan panel for about $2.00 per square foot in material costs.

Recently Edmundson has licensed Ecotechnology, Solana Beach, California, to make SolarSan panels. The price is $249 each for less than ten panels; $217 each for ten or more.

COMMERCIAL COLLECTOR PANELS AVAILABLE

Although a conscientious effort has been made to collect data about all types of solar panels, the industry is developing so rapidly that there are bound to be omissions. In the following paragraphs are brief descriptions of various kinds of panels available from small and large manufacturers. Some are partially complete and need your effort to supply the cover plate of glass or plastic. In all cases there will be some assembly work required to tie these panels into the solar system you design.

Liquid-Type Commercial Panels

The Piper Hydro System

Successfully used in quite a number of installations in California, Washington, New Mexico, and Maryland are the panels made by Piper Hydro in Anaheim, California, and usually sold as part of the company's hydronic and fan-coil system described in Chapter 4. The panel uses black copper pipes attached to blackened galvanized corrugated sheet steel in a galvanized steel tray containing fiberglass insulation. This panel is fitted with 1½-inch copper headers, top and bottom, so that it's easy to connect to the standard copper pipes used in the building industry. The cover is a single sheet of 4 mil Tedlar plastic. This panel is 8 feet long by 2 feet wide.

Sol-R-Tech Panels

Proven in New England winters, panels made by Sol-R-Tech, Charlotte, Vermont, consist of a somewhat different sandwich construction (see Figure 12). The top layer is a sheet of clear 0.040-inch-thick fiberglass-reinforced polyester. Next is 0.5 inch of air space, followed by a second infrared trap of Tedlar plastic and then another 0.5 inch of air space. Below this is a collector plate of aluminum, containing integral tubing for carrying the antifreeze mixture recommended by the manufacturer. This aluminum collector is painted with non-selective flat black paint. Below this absorber is 1 inch of fairly rigid fiberglass insulation. All these layers are sealed in a sandwich by means of a rigid aluminum "T" edge.

A useful feature of this panel, which is 8 feet long by 3 feet wide and 2 inches thick, is its light weight. It weighs only 25 pounds or just a fraction more than 1 pound per square foot. Sold by Garden Way Laboratories, an affiliate of Sol-R-Tech, this panel costs about $7.00 per square foot. It has been successfully installed on several homes and some commercial buildings in New England. Usually these panels are part of a system which includes two circulator pumps, a heat exchanger, solar thermostat, water storage tank and assorted plumbing. The panels come fully assembled.

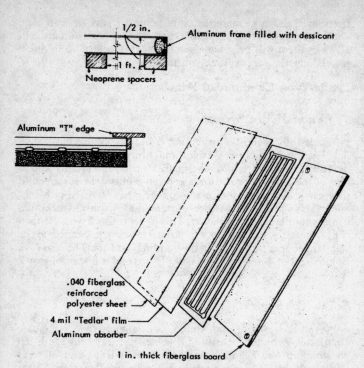

Figure 12. Construction of Garden Way solar collector.

According to Sol-R-Tech, which provides a consulting service for builders and architects as well as solar panels and some other hardware, a home in the Boston area, fitted with as few as three panels for this design, will obtain about 50% of the energy needed to heat domestic water from the sun. Materials for such a system are estimated at about $1,000.

Sunworks Liquid and Air Collectors

Another New England manufacturer of panels is Sunworks in Guilford, Connecticut. Two basic types of solar panels are provided: one uses liquid as the heat exchange medium, the other utilizes air. Both are shipped assembled, although you may purchase them without the glass cover.

There are two variations in Sunworks' liquid solar heaters. One is a surface-mounted panel readily installed on existing roofs. The other is a flush-mounted module designed to become an integral part of a roof or south-facing wall. In both cases the cover is a single plate of $\frac{3}{16}$-inch-thick tempered glass having 92% solar transmittance. Then there is an air space of about 1 inch between the cover and absorber, which consists of a copper sheet 0.010-inch thick to which are soldered ¼-inch copper tubes ⅜ inch, outer diameter (OD), spaced 6 inches apart. The headers or manifolds are made of 1-inch copper pipe (1.125 inch OD) and the longitudinal tubing is soldered to the headers with silver solder.

One feature considered important by Everett Barber, president of Sunworks, is the use of a selective black coating on the copper absorber sheet, tubes, and headers. The coating he recommends is made by Enthone, New Haven, Connecticut. It has a minimum absorptivity of 90% and a maximum emissivity of 12% of the incident solar radiation. Selective black finishes are more expensive than ordinary flat black paint but their advantage is that much less heat is reflected from the absorber surface.

Below the Sunworks absorber is an insulating pad of 2½-inch-thick fiberglass. The entire sandwich is contained in a box made of aluminum sheet at the bottom and aluminum extrusions (built-in channels) on the sides. A neoprene gasket and aluminum extrusions holding the cover glass in place help to make this panel weathertight. This panel is 7 feet long, 3 feet wide and 4 inches thick with an effective absorber area of 18.56 square feet. Price is $10.57 per square foot in quantities less than eleven. Weight of the panel is 120 pounds when filled with antifreeze.

Construction of the Sunworks panel using air as a medium is quite similar in its choice of materials. Instead of copper tubing to carry liquid, there are eight fins, or vertical metal plates, each 2 inches high and bonded to the back of the selective black absorber plate of copper. Passage of air is between the fins, from top to bottom of the panel, behind the absorber. Air is brought in through a 2-inch by 6-inch duct stub—to which the building's ductwork is connected—and exhausted through a similar duct stub at the bottom.

Dimensions of this air heater are 7 feet 10½ inches long by 2 feet 9 inches wide, and its effective absorber area is 19.1 square feet. Weight of the module is 110 pounds and recommended air flow through the panel is at the rate of 3 cubic feet per minute for each square foot of collector area.

If you're interested in building a solar panel of the Sunworks liquid type yourself, the company will furnish the absorber only.

That is, you get the copper absorber plate and tubing coated with selective black paint at a price of $6.00 per square foot for a small quantity.

On a clear day, Barber estimates that his Sunworks panels can produce temperatures of 100° to 150° higher than the ambient. So on a hot summer afternoon, you can produce temperatures higher than the boiling point of water with the liquid-type panel. If there is no flow of liquid, this kind of panel can reach a temperature of 400° F.

Tranter Quilted Solar Panel of Steel

A different variety of liquid-carrying flat plate collector is made by Tranter in Lansing, Michigan. This is a panel with a quilted weld pattern. It is made of either carbon steel or stainless steel, with a considerable differential in cost. These Econocoil solar panels are available in four models, two having an area of 10 square feet and the other two, an area of 20 square feet. Without any cover to trap reflected radiation, or insulation below the absorber plate, the unpainted carbon steel versions cost about $8.00 per square foot. The stainless steel models range from $20 to almost $27 per square foot.

When tested by NASA in Cleveland recently, these Econocoil plates were covered with two layers of ⅛-inch glass having a transmittance of 88%. The absorber was painted with a flat black nonselective paint.

Water flow rate was 10 pounds per hour per square foot of collector. At an inlet water temperature of 80° F and an ambient temperature at the same level, efficiency of this Tranter system was calculated at 70% in collecting solar energy. This amounted to a gain of 210 British thermal units (Btu) per hour per square foot of collector.

When the temperature of incoming water reached 200° F, the efficiency of the Econocoil unit dropped to 38%. This is to be expected because the relative amount of solar energy absorbed by a hotter surface will always be less than that of a cooler one.

According to the manufacturer, this collector plate may be reduced in price if mass production is achieved. This is certainly desirable if such units are to be competitive, because the price, without glazing, painting, or a mounting tray and associated hardware is higher than many complete collector panels.

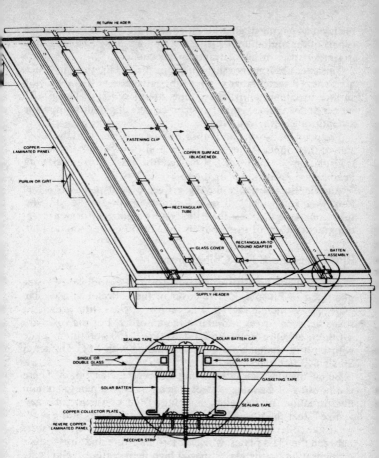

RETURN HEADER

COPPER LAMINATED PANEL

PURLIN OR GIRT

FASTENING CLIP

COPPER SURFACE (BLACKENED)

RECTANGULAR TUBE

GLASS COVER

RECTANGULAR-TO-ROUND ADAPTER

BATTEN ASSEMBLY

SUPPLY HEADER

SEALING TAPE

SOLAR BATTEN CAP

SINGLE OR DOUBLE GLASS

GLASS SPACER

SOLAR BATTEN

GASKETING TAPE

COPPER COLLECTOR PLATE

SEALING TAPE

REVERE COPPER LAMINATED PANEL

RECEIVER STRIP

SOLAR BATTEN

Figure 13. Integral roof collector panel of copper, designed by Revere.

Revere's Copper Collectors

The liquid-type solar collector designed and marketed by Revere Copper and Brass can be used as part of the roof of a building since it is designed for flush mounting. Details of this panel are shown in Figure 13.

A typical Revere collector comes fully assembled. It is 6½ feet long and 3 feet wide, with a depth of 5½ inches. Its cover is a double layer of tempered glass mounted in an aluminum frame which is part of the aluminum housing for the solar panel. This housing is laminated to an inner frame of plywood.

The solar absorber is a sheet of copper 0.016 inch thick to which are fastened rectangular copper tubes attached to the absorbing surface by copper clips. These tubes are placed at distances of 8 inches or 5½ inches, from center to center, depending upon the amount of heating required.

The copper absorber and tubes are painted black, and the lower side of the copper is bonded to ⅜-inch-thick plywood. Below this is insulation, 3½ inches of foil-backed fiberglass. Top and bottom headers are ¾-inch copper tubes.

Another standard size for these Revere solar panels is 8 feet long by 2 feet wide, and includes from two to five rectangular tubes on the solar absorber, depending upon performance requirements. Cost of these panels is about $7.60 per square foot.

PPG Baseline Solar Collectors

A second giant producing solar panels is PPG Industries, headquartered in Pittsburgh. The cover of the collector is a double glazing of ⅛-inch thick Herculite-tempered glass with ⅜-inch air space between layers and above the absorber. For the solar absorption, an Olin-Brass Roll Bond Type 1100 aluminum plate with built-in channels for liquid is painted with PPG Duracron Super 600 flat black coating.

Below the absorber is 2½ inches of fiberglass insulation. There is a housing of galvanized steel which serves as a pan to contain the insulation; and an edge retaining system consisting of a galvanized steel channel, which holds in place the edges of the sandwich-like collector elements. PPG also uses desiccant-type spacers between the glass plates, which absorb moisture and thus reduce condensation on the glass. Special high temperature sealants are also used at the panel edges to withstand temperatures which may run as high as 400° F when there is no liquid in the panel. Preferred as the fluid is a mixture of distilled water and ethylene glycol in areas where the collector panel will experience freezing temperatures. Typically a 50% mixture of such an antifreeze will lower the fluid's freezing point to −34° F. In any case, the pH of the mixture should be between 6 and 7 to avoid corrosion of the aluminum absorber pipes.

Dimensions of the PPG panel are 34³⁄₁₆ inches by 76³⁄₁₆ inches by 1⁵⁄₁₆ inches. Containing either pure distilled water or an antifreeze mixture, the collector weighs about 6 pounds per square foot. They come assembled and cost about $7.00 per square foot. These panels have been extensively tested in Florida and Pennsylvania. They are also installed on an experimental solar home built by Ohio State University and Homewood Corporation at Columbus, Ohio. Other installations are planned in the near future by builders such as Homewood, on houses ranging in price from $30,000 to $50,000. Appearance of these panels, like most other solar collectors, detracts nothing from the appearance of the house when they are correctly installed.

Reynolds Aluminum Panels, Units, Kits and Parts

About two years ago D. J. Laudig, manager of the Reynolds Metal plant in Torrance, California, became interested in making solar panels. With the sanction of top management in Richmond, Virginia, he and his associates designed a collector made largely of aluminum.

Choosing dimensions of 8 feet by 4 feet as a convenient size for new construction, the Reynolds team designed a sandwich for collecting sunlight efficiently. They named the alloy used in this panel Solarum. Aluminum tubes having an ID of 0.420 inch and OD of 0.520 inch are welded to the extruded absorber surface. The eight tubes are connected to top and bottom headers inside the extruded frame, in such a way that there is a single inlet and a single outlet tube. This makes for a minimum of plumbing connections. The absorber plate and tubes are painted with a flat black finish.

Underneath the absorber is 1½ inches of Celotex insulation. The cover for this Reynolds panel is not supplied by the manufacturer, since it's easy to attach a cover during installation of the panel. Clips are supplied with the frame for this purpose, and are designed to hold a bottom glaze of Tedlar—used in several buildings where these panels are in service—and a top sheet of clear fiberglass (allowing for an air space in-between).

At present, Reynolds Metals, just entering the solar field, is offering their Model 14 Solarum panel in almost any form you want: as an assembled unit; in kit form; or in parts. For instance, if you want to buy only the absorber plate, including tubing or other parts, this big manufacturer is very accommodating.

An assembled panel 8 feet by 4 feet by 3⅝ inches weighs 64 pounds dry and 67.8 pounds with the recommended antifreeze mixture as the working fluid. This is just over 2 pounds per square foot when installed. With a solar collector area of 29.5 square feet, this panel absorbs between 30,000 and 50,000 Btu per day in tests in Southern California. Price of an assembled panel, without cover, is $256 or $5.00 per square foot, a highly competitive price.

Reynolds installations

Installations of the Reynolds panel include a roof at Sambo's restaurant in Calabasas, California, where the solar system installed by Elsters, a Los Angeles engineering firm, supplies heat for hot water used by a busy operation. In this case, Reynolds has also supplied an aluminum heat exchanger costing $25. It is used to transfer heat from the sun-warmed antifreeze to the local water.

Another installation is on the roof of a home designed and owned by Charles L. Martin in Palm Desert, California, where solar energy heats both domestic hot water and the swimming pool. A similar application is on the roof of a 6,000-square-foot home in Tucson, Arizona, built by Environmental Aire. This installation expects to extend the use of solar energy to provide interior heating and cooling, as well as heating domestic water and the pool.

Corning Tubular Panels

Another large manufacturer, Corning Glass Works, Corning, New York, has developed a novel and highly efficient type of solar collector shown in Figure 14. In this example of the unit are twelve glass cylinders, evacuated like fluorescent lamps. Each cylinder is 10 feet long and 4 inches in diameter.

Inside each cylinder is an absorber plate made of copper and painted with a selective black coating. The same type of coating is applied to the attached copper tubing containing the heat transfer fluid, which is water. In cases where an antifreeze mixture or water containing a corrosion inhibitor is used, Corning may supply collectors using steel or aluminum tubing with some reduction in cost; but copper is the only metal used in the panels currently available.

According to the manufacturer, these panels are from two to three times as efficient as conventional two-pane flat plate collectors in terms of energy delivered per unit of absorber area. Working fluid temperatures of 250° to 300° F are achieved with good efficiency.

Reasons given by Corning for the added efficiency of their panels include: lower optical loss, since the transmittance of the glass tubing is 92%; lower thermal loss, because both the absorber plate

Figure 14. A rack with Corning glass tubular solar collectors.

and the attached copper tubing carrying the water are in a high-vacuum tube and are thereby well insulated against wind and cold weather; and an extremely selective, non-radiating absorber.

According to Corning managers, it is feasible to mount these new evacuated collector tubes without any external insulation. They come fully assembled and can be mounted in simple racks. Copper tubing carrying the solar-heated water is brought out of the high vacuum glass tubes through glass-to-metal seals proved in years of service.

Because Corning's design permits considerable versatility in the size of this solar panel, it is being offered to architects as a collector system that can be mounted in a variety of ways: horizontally on a flat roof; at an angle on a tilted rack or sloping roof; or vertically on a south wall. The firm is furnishing modular panels having from six to ten tubes, each tube 8 feet long or more. While so far the tubular collectors have been made in lengths no greater than 10 feet, Corning engineers state that they appear to be structurally feasible in lengths up to 20 feet.

Since these panels have high efficiency and module size may be varied, it's possible to install them on vertical walls and flat roofs as well as on sloping roofs with relatively minor loss in the absorption of the solar energy. Although the tubes are mounted in racks without installation, manifold or header piping connections must be well insulated to avoid loss of solar energy as the superheated water passes through them.

Corning installations

In tests made in various United States locations, these Corning panels have performed well. One of the first installations is on the roof of Colorado State University Solar House III at Fort Collins.

In this Colorado application, the Corning modules are in arrays 8 feet long by 4 feet wide. Because of the climate, the heat transfer fluid is a mixture of water and antifreeze. Tests show that when this liquid is at 200° F, efficiency of the collector panel is as high as 70%. When the liquid temperature is 350° F, collector efficiency of 20% is achieved. The super-heated fluid transfers energy by means of a heat exchanger to a storage tank containing 1,100 gallons of water. The tank is insulated with 8 inches of fiberglass, truly heavy insulation. In the summer, this house is cooled by a 3-ton Arkla lithium bromide absorption cooler powered either by hot water from the storage tank or the hot antifreeze mixture from the collector, since the valving of the delivery system to the cooler permits a choice of fluids. When the collector alone is used, fluid can be delivered to the absorption unit at temperatures up to 350° F.

Construction of this experimental house was funded by Colorado State University, while money for the continuing performance study has been granted by the National Science Foundation.

Six Other Manufacturers of Solar Panels

Here is a summary of designs by additional manufacturers of solar panels. All the following panels come completely assembled. Complete addresses can be found in Appendix I.

SOLAR POWER CORPORATION, New Port Richey, Florida, makes the TK 200 solar collector, available with several options as to glazing, and with several accessories for space and water heating. The basic panel contains a copper absorber plate and copper piping, all painted black, in an insulated redwood container. This is furnished in several ways: either without glass (but provisions for you to add either single or double glazing of ⅛-inch panes); with single glass cover; with double glass cover; or with the Kalwall Sun-Lite fiberglass-reinforced transparent sheets in double glazing.

Dimensions of these panels are slightly over 8 feet long by 2 feet 2 inches wide. Weight is about 78 pounds or nearly 5 pounds per square foot. Prices of the panels range from about $14 per square foot without glazing to about $18.50 with the double plastic glazing. Accessories include a valve package and choice of two pumps, one with a differential thermostat control unit.

CSI SOLAR SYSTEMS DIVISION, St. Petersburg, Florida, makes copper collectors and supplies a small pump to permit you to tie their solar system into existing plumbing for heating domestic water. Their Sol-Heet system has been found capable of supplying an average family's hot water requirements through three days of sunless weather. The pump costs about $.30 a month to operate; solar energy is free.

E & K SERVICE COMPANY, Bothell, Washington, makes 4-foot by 2-foot liquid-containing panels with a glass cover. This relatively small panel costs $60 in quantities of ten to twenty-five units, each panel having a capacity of 20 gallons of fluid. A complete system for space and water heating includes a Rho Sigma differential thermostat, two pumps, a large water storage tank and an auxiliary water heater fired by gas, oil or electricity for standby service in prolonged bad weather. In a test in Denver, two panels heated 20 gallons of water from an uninsulated tank from 47° to 83° in three hours on a clear day, when the ambient temperature rose from 40° to 55° F.

ENERGEX CORPORATION, Las Vegas, Nevada, makes several models of liquid-type copper solar panels for various heating purposes.

HITACHI AMERICA, New York, New York, makes a solar collector for hot water consisting of six plastic cylinders and a plastic cover contained in a steel box. See details in Chapter 5 on Hitachi hot water systems.

SUNSOURCE, a subsidiary of Daylin, in Beverly Hills, California, is offering a complete water heating system based on a solar collector panel designed by the Miramit Company of Israel. One of the features of the Sunsource panel is its use of a special selective black absorber coating that has been proved effective in hundreds of solar water heating installations in Israel. This selective coating, with an absorption capability of 92% of incident insolation, is applied to a galvanized steel sheet above which is either a single or double glazing of double-strength glass. Below the absorber plate is 1½ inches of rock wool insulation. Compression-bonded to the plate are seven galvanized steel pipes, ½-inch inner diameter (ID); 1-inch ID pipes of the same material are used as upper and lower self-manifolding headers, that is, the long head-

55

ers come already built with attachment for the panel outlets. The entire sandwich is contained in a tray made of 24-gauge galvanized steel, with silicone sealant around the glazing. Dimensions of the Sunsource panel are about 6 feet long by 3 feet wide with a depth of 3½ inches. Weight of this panel, without water, is 134 pounds including a single layer of glass, two panes each 35 inches by 35 inches.

With this type of panel, water temperatures obtained by tests in Southern California will reach between 180° and 200° F. According to Sunsource, it is possible to build a supplementary hot water system for your home, using two collector panels, for less than $700 in cost of hardware.

Imported directly from Israel are panels made of the same materials as the Sunsource unit, including galvanized steel for absorber plate, pipes and container, plus rock wool insulation. These imported collector panels are made by Amcor and distributed by SOL-THERM CORPORATION, New York City. A panel is 6 feet long by 3 feet wide and weighs 120 pounds. It's priced at $245 in a quantity of four panels or fewer. Sol-Therm offers a package which includes: a solar water heating system with two collectors, a 32-gallon storage tank, all necessary pipes and nipples for connecting the tank to the collectors, and a mounting frame so that you can install the collector panels on a flat roof. Price of this package is $695.

SUNWATER COMPANY, 1112 Pioneer Way, El Cajon, California, makes collectors 10 feet long by 3 feet wide using aluminum, coated with black silicone rubber paint as the absorber. Liquid is the heat transfer medium, passing through copper tubes bonded to the back of the absorber with silicone rubber bonding. Below the absorber is a 2-inch insulation of fiberglass board. Above the absorber is a double glazing of glass, with ¼-inch air space between the layers. Size of each pane is kept to 3 feet wide by 1½ feet long, and they are held in position by sheet steel framing.

According to Sunwater officials, the Sunway system, properly installed, will raise the temperature of 1½ gallons of water by 65° F per square foot of solar panel on a clear day. So it's very easy to calculate how many panels are needed for any given installation.

In addition to 10-foot by 3-foot solar panels, Sunway is now offering 8-foot by 4-foot units, as this latter size is becoming somewhat of a standard for solar heating in the building industry. Price of the 10-foot by 3-foot Sunway solar panel is $6.66 per square foot to builders and contractors.

Air-Type Commercial Collectors

More fully described in Chapter 4 are two types of collector panels using air as the transfer medium, and both developed in Colorado. Perhaps this is not too surprising in view of the fact that Dr. Löf's solar homes in Denver and Boulder—the oldest continuously occupied solar houses in the United States—include solar collector panels with an air transfer system.

Solaron Air-Type Collectors

In the system made by Solaron Corporation, Denver, there are two sheets of double-strength window glass spaced ¾ inch apart for glazing. The solar radiation passing through this is absorbed by a galvanized steel sheet painted with non-selective black. Air is blown through a ½-inch space between the steel absorber and the insulated roof into ductwork for heating the house or for storing the solar energy in an insulated bin full of dry stones.

From the lower end of this solar panel to the upper end, there typically will be a rise in temperature of around 50° to 75° F. Cool air is returned automatically to the collectors for reheating by a blower system which is controlled by a thermostat. To heat water, there is an air-to-water preheater always in series with the roof collectors—that is, they are in the same continuous line of piping. During periods when there isn't adequate sunshine, an auxiliary heating system takes over.

It's estimated that in a relatively cold area such as Denver, a Solaron system will provide 75% of the load in heating a home and its hot water. Storage with a 15-ton bin full of stones will carry the house through for one day in January when the outside temperature is −10° F. When the ambient is higher, the carrythrough from stored solar heat is longer.

International Solarthermics System for Retrofitting

A small research company in a town with a population of five hundred has created a useful type of solar panel and reflector, with air as the transfer medium to a storage bin right next to the collector. International Solarthermics Corporation (ISC), Nederland, Colorado, has designed a complete solar furnace for heating homes. It is the A-frame solar furnace briefly described. Collectors are made of aluminum sheet painted non-selective black, on which are mounted vertical vanes, small, cup-like structures also of black-coated aluminum.

The collector covers are a double layer of $\frac{3}{16}$-inch float glass—inexpensive but clear—and air is blown between the glazing and the collector plate to the stone storage bin. This bin, as well as the back of the aluminum collector plate, is heavily insulated with fiberglass.

The absorber plate's vanes might be described as a large number of small cylinders, each 2.75 inches in diameter and 2 inches high. Made of the same 0.019-inch-thick aluminum as the plate to which they are fastened, these vanes increase the black absorbing surface area of the total assembly. As the air flows in and out of these cups, blown from the bottom of the A-frame panels shown in Figure 15, it absorbs more heat than passing over a flat black sheet.

Another important feature of the ISC design is the reflective shield. For economy it is made of highly reflective aluminum foil bonded to a plywood backing. This reflector has a dual purpose. During the entire heating season, it reflects sufficient solar energy to the collectors so as to improve their efficiency by about 30%. When the solar furnace is not in use, the hinged reflectors are raised like lids to cover and protect the glass of the collector panels.

There are three standard sizes of collectors with areas of 96, 128 and 160 square feet depending upon the size of house to be heated. Corresponding reflector areas are somewhat larger, being 108, 144 and 180 square feet.

There are some other interesting aspects of this air-type collector. First, it is built into a system called a solar furnace, and not

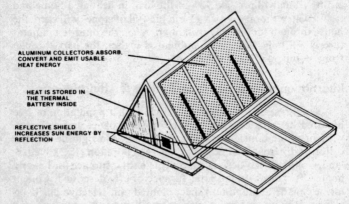

ALUMINUM COLLECTORS ABSORB,
CONVERT AND EMIT USABLE
HEAT ENERGY

HEAT IS STORED IN
THE THERMAL
BATTERY INSIDE

REFLECTIVE SHIELD
INCREASES SUN ENERGY BY
REFLECTION

Figure 15. A-frame shed with air-type solar collectors by International Solarthermics.

marketed as a separate collector panel, although considered solely as a collector, it's a highly efficient design. Then, the company that designed it is not manufacturing it. Instead, ISC has already licensed seven manufacturers to build, distribute and in some cases, install these solar furnaces.

Estimates of ISC officials are that more than thirty-five hundred homes will be equipped with this solar heater before the end of 1975. This is because this solar furnace can readily be installed as a retrofit, attached directly to the forced-air heating system of existing homes.

4 / A Solar Furnace to Heat Your Home

In the preceding chapter we've seen how easy it is for a handy person to build a solar collector; we also reviewed the many types of air and liquid collectors available either in kit form or fully assembled.

The next step is to apply this collected solar energy. This chapter will discuss how to build an air or liquid solar system to heat your home. But first you might like some answers to a few basic questions.

EFFICIENCY OF SOLAR HEATING

How Much Heat Will a Solar Collector Produce?

The amount of heat you can get from a solar collector will vary depending upon the following factors:

As we saw in the last chapters, the materials used in fabrication of the collector and its design affect efficiency. Some designs are definitely superior to others for home heating, partly because of a better choice of materials. For instance, collectors made of plastic are not as good absorbers of heat as those where blackened metal is the absorber. Therefore, plastic is generally used only for heating swimming pools and domestic hot water, in applications where its lower initial cost is appealing.

As ambient temperature decreases, collector efficiency will decrease. There's a linear relationship between the temperature difference of the outside air and the outlet of your solar collector. If

$$T_{difference} = T_{outlet} - T_{air} \text{ in degrees F}$$

then the following brief tabulation will illustrate this point, assuming an average value of 250 Btu per square foot per hour of solar energy striking the surface of your collector panel.

Collector Efficiency (%)	$T_{difference}$ (°F)
70	0
60	32
50	55
40	90
30	125
30	125
20	145

Table 1. Ambient temperature effect on collector efficiency

Solar intensity, or the amount of insolation on your collector, is also important. Anything that cuts down the amount of sunlight, and hence heat energy, reaching your absorber will adversely affect performance. Smog, haze, clouds, fog and rain, as well as dirty or shaded cover panels, reduce efficiency.

Orientation of the collector makes a considerable difference in its performance, as pointed out in Chapter 2. For best results in the Northern Hemisphere, your panel surface should face about 10° west of true south. A collector facing due east or due west is only about 40% as efficient. Inclination of the panel should be at an angle of about 10° greater than the latitude of your area, as also explained in Chapter 2.

How Much Energy is Required to Heat a House?

Evidently the answer to this question will depend on many factors: how big your house is; how well it is insulated; its location as it affects weather and wind.

There are some detailed analyses of these and other factors available in such publications as the *ASHRAE Handbook: 1974 Applications Volume* published by the American Society of Heating, Refrigerating and Air Conditioning Engineers, and *Design Criteria for Solar-Heated Buildings* by Everett H. Barber, Jr. and Donald Watson.

For a three-bedroom house that is well insulated, the annual requirements for heating alone can range from less than 50 million Btu to 150 million Btu. This is one reason why, in a house without summer cooling, about 80% of the total energy requirements are for heating. When you realize what a large percentage of your annual utility bill is paid to heat your home, solar heating becomes increasingly attractive.

As indicated earlier, you can expect to get anywhere from about 50% of your space heating needs to a full 100%, depending upon the location and insulation of your house, and the size and efficiency of your solar heater. A good average appears to be 65% to 75%

CHOOSING A SOLAR HEATER

Now is the time to renew the argument between air-type and liquid-type solar heating systems. In previous chapters you've learned some of the basic advantages of each type and the general principles of construction. In this chapter are some details as to how you can build each type of solar heating system and apply it to your home.

Here are some general considerations in making your choice between an air-type and a liquid-type solar furnace.

Planning Around an Existing Heating System

If you want to apply a solar heating system to your present home, and it already has a forced-air heating system, you'll probably find it easier to install an air-type solar furnace. This is particularly true if your roof does not have a south-facing area large enough for installing solar collector panels.

If your house has a radiant heating system of the hydronic type, with hot water pipes in your floors, baseboards or ceilings, then a liquid-type solar collector system is likely to be your choice. You can tie the solar-heated liquid into your present system either through a heat exchanger or, if you don't have to worry about freezing weather and use water in your collectors, you can bring the solar-heated water directly into your present hot water heater.

Planning Around Your Heating Needs

If you want your solar system to handle both space heating and domestic hot water, it may be a difficult choice between air and liquid solar systems. It is easier to design a system for both purposes with liquid as the heat transfer medium and will probably be somewhat less expensive if you can do some of the work yourself. However, if your home is now equipped with a forced-air heating system, there are a couple of methods for heating your hot water with an air-type system, described later in this chapter, that may make an air solar furnace preferable.

If you expect your solar system to handle cooling as well as heating the rooms and your hot water, your choice will be influenced by your location. In areas where you can count on pumping cool air into your home at night, or where you can bring it in after midnight at ground level through a small air conditioner, an air system will probably be most economical. However, a highly efficient array of liquid collector panels, furnishing superheated water or antifreeze mixture at temperatures up to 300° F, can provide the energy for air-conditioning and will have some advantages in many areas where summer heat is a real problem.

For the heating and cooling application, there is a third choice most useful in hot dry areas. It's the Skytherm house designed by Harold Hay and Associates and described in Chapter 6. In this design, large black plastic water bags on a flat roof provide winter heating and summer cooling.

BUILDING AN AIR-TYPE SOLAR HEATING SYSTEM

It will be easier to consider how you build your air solar heater if we start by assuming that your house has a forced-air heating system. Here's how to go about it.

Where to Place Equipment

Determine whether you want solar collector panels on your roof or not. Your decision on this matter will depend on two factors: whether you have adequate space on a south-facing roof with a slant from the horizontal 10° greater than your latitude; and whether you can design a way to connect ductwork from your present heating system so that you can blow air in a ½-inch space between your solar panels and the roof. This air must travel from the lower edge of your roof to a duct at the upper end of the roof. This latter duct must bring hot air into your present ductwork. (See Figure 4.)

Another decision you must make is where to locate the large bin of small rocks or pebbles for storing the solar heat. If you have a basement that once had a coal bin, this would be a good solution. You will need a space that can be thoroughly insulated, will hold about 15 tons of small, clean stones, and through which you can blow air from and to your present heating system.

Making Air-Type Collectors

Let's assume that everything is favorable and you decide you want air-type solar collectors on your roof. The next step is to

Figure 16. Air-type collector panels have heated Löf house in Boulder, Colorado, since 1944.

build or buy your air-type solar collector panels. (See Figure 16.) The techniques described in Chapter 3 for building collectors apply, except that you're handling air instead of liquid.

For a convenient size, build your panel with an absorber made of a corrugated sheet of galvanized steel—ordinary roofing material—8 feet long by 2 feet wide. Paint the upper side of this with a selective black paint. A first coat of Rust-Oleum black graphite, a good heat conductor, and a second coat of 3M No. 101-C10 Black Velvet, a selective black for excellent absorption, will do the job nicely.

Although many use two layers of glass, our recommendation for simplicity, durability, and low cost, as well as lighter weight, is to use a single cover sheet of .040-inch-thick fiberglass-reinforced polyester such as the transparent material made by Kalwall. This is being used successfully by several manufacturers of solar collectors.

Building a Collector Frame

Next, make a rectangular frame of painted galvanized steel. This frame must be wide enough to hold as many collector panels as necessary to meet your heating needs. This means that if, say, you need ten panels, it will be slightly more than 20 feet wide and

8 feet long at the sides so as to contain ten 8-foot by 2-foot panels. This frame will also have grooved cross members of galvanized steel at the top, spaced at 2-foot intervals, to hold the plastic cover sheets.

At the lower end of this frame you need a ductwork input from your home heating system. A wide but relatively shallow duct, gasketed into the lower end of your steel frame, will do. The connection from your home forced-air heating system will depend on the most convenient outlet. A neighboring sheet metal or plumbing supply store can give you good advice as to available ducts, both rigid and flexible. You may want to use plastic ductwork for at least part of your connections.

A similar output duct connection must be made at the upper end of your array of air-type connector panels. This output must lead as directly as possible through a duct system to your rock storage bin. Depending on the distance to your stone bin, you may need to insert a blower so that the hot air from the collectors is forced into the storage bin.

Constructing a Rock Bin for Heat Storage

The bin for storing 15 tons of pebbles should be well insulated. If you do not have a basement, find some other area large enough on the first floor of your home. A good way to build the storage bin is to make it of concrete, then line it with at least 4 inches of insulation such as fiberglass or rock wool, and inside that, a wooden bin. Remember that there must be a duct to bring the hot air in from the collectors, and another duct with an exhaust fan to force the heated air out—into your home and the return system that eventually carries it back to the solar collector.

One other consideration in building your rock bin is whether you want to use this solar heat to provide you with hot water. If so, one way to do it is to place a hot water tank inside the rocks. This tank can be connected by piping to your present hot water heater and thus serve as a preheater. It will remove some of the load from your conventional water heater and consequently save you additional money. For this system you will need a pump and differential thermostat, about which more will be said later in this chapter. The water in your solar-heated tank should feed in from your normal water supply, and be fed out when heated to the top of the tank.

What you've accomplished by placing a water tank in your stone bin is to build a large air-to-water heat exchanger. The tank should stand upright in the bin because the upper part of the bin will be hottest. Supply water is brought into this tank at the bottom and taken out at the top.

Larger bin required for water heating

Modifying your rock bin to include a water tank means building a larger bin than you would if you used this heat storage area solely for heating your house. You should allow for a size increase of at least 25%, and preferably more in the tonnage of heated stones. Thus your bin capacity might hold 20 tons of pebbles plus the preheat water tank.

Another important consideration is the stones themselves. Ideally these pebbles should be the size of a small bird's egg, smooth and round—river gravel is good. You should be sure to wash these pebbles and get them thoroughly clean before putting them in your bin. Otherwise, your solar heating system will blow dust all over your house.

AIR HEATING SYSTEMS FROM MANUFACTURERS

Several companies are making air-type solar collector panels and associated hardware, including control systems. There is a list of several, including Kalwall, Solaron and Sunworks, in Appendix 1.

In most cases these manufacturers provide a consulting service for architects and builders who want to install air-type solar furnaces. Also some of them have instruction manuals for sale. These booklets can be helpful in your choice of equipment as well as in the installation of your solar system.

International Solarthermics A-Frame Shed

With an existing hot air system in your present home, it is certainly worth trying to place your solar furnace in a suitable location on the ground in your yard.

One approach previously mentioned looks particularly good. It is the International Solarthermics Corporation (ISC) A-frame shed, which includes solar collector and reflector panels, a rock storage bin, blowers, ductwork and a control system—a complete air-type solar furnace in a building no larger than a tool shed.

The ISC collector panels were previously described. Their unusual cup-like vanes on the aluminum absorber panel have the effect of reducing both reflective and radiative losses, as well as of minimizing convective and conductive losses. Also, there is a larger area for absorbing solar energy than that of a flat panel, but the space occupied isn't increased—only the thickness of the panels, by the 2-inch height of the cylindrical vanes.

One other effect of the vanes is to interfere with the high-speed air stream from the fan in the A-frame, shown in Figure 17, so that there is efficient heat transfer. While you're operating this system, the air stream is at about the same temperature as the absorber plate.

Sealing of the entire structure is accomplished with high-temperature, water-based acrylic caulking, which has the kind of "memory" needed to compensate for expansion and contraction of large panels of cover glass.

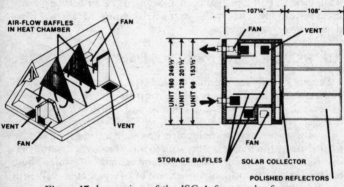

Figure 17. Inner view of the ISC A-frame solar furnace.

Reflector Improves Efficiency

Adding about 30% to the efficiency of the ISC collector panels is the reflector shield, bright aluminum foil bonded to plywood, and hinged to the collector so that it can be closed and used as a protective cover during summer months when the solar furnace is not in use. (See Figure 18.) The 30% gain in energy absorbed by the collector with the use of this reflector is based on a measurement where the collector is at an angle of 60° with the ground at a latitude of 40° North.

Of course, a major advantage of the A-frame construction is that you can select the angle of collector inclination that is most effective for the latitude of your home.

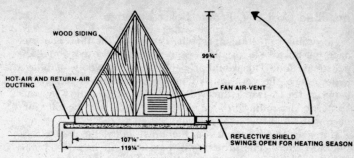

Figure 18. Front layout of ISC A-frame solar furnace.

Efficient Heat Storage

A fan contained inside the A-frame blows air from the outside through the space between the collector and the glazing. The heated air from the vertical collector passes through internal baffles and into the rock storage bin. Stones ranging from ¾ inch to 1.5 inches, such as crushed granite, limestone or sandstone river gravel, are stored in a compartment from floor to peak of the A-frame shed. Three furnace sizes are available, and amounts of stone used range from about 11 to 22 tons. This type of bin has been found to have a practical heat storage capability from 75° to 175° F, and to be able to provide carrythrough heating for three to five days without useful sunshine. Heat loss from the bin rarely exceeds 3° F during any twenty-four hours.

Behind the collector plate is heavy insulation consisting of laminated ⅜-inch plywood, 4 inches of polyurethane foam, and ⅜-inch insulating board; the same insulation is used for the opposite wall. The base of the A-frame uses laminated ½-inch plywood, 6 inches of polystyrene foam and ⅜-inch insulating board. Below that is a foundation having 6 inches of polystyrene foam with 0.006-inch polyethylene vapor barrier. Framing the structure is 24-gauge galvanized steel.

From the rock storage bin a second fan exhausts hot air through conventional ductwork to the house. ISC specifies two Lau Industries model FGP10-6A centrifugal or radial-flow fans powered by two GE motors of the ½-horsepower permanent-split capacitor type.

Installed Cost of A-Frame Solar Furnace

According to the engineers who developed the ISC solar furnace, it is possible to install one of their units at a cost of about $20 per square foot. This would probably mean that you would have to do some of the work yourself, including making the connection between the A-frame ductwork and the nearest duct in your home. So, depending on what panel size you choose, the solar furnace will cost between $2,000 and $3,200.

If it saves $600 a year in your fuel bills—and you may save considerably more—this furnace will pay for itself in a few years. After that, you'll be paying a good part of the taxes on your home by the money saved in fuel bills.

Heating Water with an Air-Type System

Although the designers of the ISC solar furnace haven't suggested in their literature that their system could also be used as a supplement for domestic water heating, all you need to do is look at the system designed by Solaron, another Colorado company, and see that it is feasible. This was illustrated in Figure 4.

Merely by including an air-to-water heat exchanger as part of your home ductwork system, you can use some of the solar heat for preheating water for your regular hot water tank. That means that there will be less use of gas, oil, or electricity for heating your water.

Designing an air-to-water heat exchanger

You might design such an air-to-water heat exchanger by wrapping about thirty turns of flexible copper tubing—about 60 feet—around the metal duct inside your house which is connected to the outdoor solar furnace. The input end of this tubing will come from your water supply; the output will be fed into the upper end of your conventional hot water heater.

The purpose of this heat exchanger is to act as a preheater for your water and thus save you fuel. Of course, the cold water coming from your supply will be drawing some of the heat energy generated by your solar furnace, so that you may need a larger size than originally planned for. As noted before, the three standard ISC collector sizes are 96, 128 and 160 square feet.

These collector dimensions give a clue to the heating efficiency of this system, since most liquid-type solar heaters require a somewhat greater collector area.

If you use an air-to-water heat exchanger of the kind just described, be sure to wrap the copper coils around the duct with a heavy layer of insulation. Typically at least 3 inches of glass wool held in place with good plastic tape will do the job.

All ductwork throughout your air heating system should be well insulated as well as the doors, windows, walls and roof of your house. Every maker of solar heating and cooling equipment emphasizes the value of good home insulation. If your home is not truly well insulated, before you install a solar furnace, consider that spending $1,000 in improved insulation will probably save you $1,500 or more in the cost of your solar system.

Every solar home we've studied is well insulated. Otherwise the owner would have been wasting much of his money in putting in large-scale solar heating.

BUILDING A LIQUID-TYPE SOLAR SYSTEM TO HEAT YOUR HOME

In Chapter 3, you learned how to build liquid solar collector panels. In the next few pages are instructions of piping arrangements in a solar heater, as well as details for installing collectors on a roof. (See page 98 for updated information.)

Although this piping scheme (Figures 19-30) was prepared by Sol-R-Tech for use with their panels and is reproduced with the manufacturer's permission, the same principles apply to most types of liquid solar collectors.

Antifreeze and Water as a Heat Transfer Medium

Although water is often used as both the heat transfer and storage medium, if you're building a solar heater in an area where there's freezing weather, you'll probably want to use an antifreeze mixture (50% ethylene glycol, 50% water typically) in your collector panels and insert a heat exchanger into the system.

However, if you prefer to use water throughout there are two alternatives. You can drain your collectors every night as Thomason and Shore do. Of course, they're using only trickling water, not the flow common to most liquid collectors. If you drain, there's always the possibility of residual water turning into ice and damaging piping or collectors. Or, you can keep the top of your storage tank warm with the auxiliary heater if the temperature falls below 160° F (average water temperature should be 140° F). You can keep your heat pump running at night or during a long cloudy spell; but this is not economical use of your heat pump or water pump.

Therefore, if freezing weather is usual in your area, our recommendation is to use antifreeze and include a heat exchanger in your system. As mentioned in Chapter 3, one source of such a unit is Reynolds Metals, which supplied a heat exchanger made of coils of aluminum piping in an aluminum container at a price of $25 to users of the Reynolds collector panels.

One other precaution, if you're using an aluminum collector, is to be careful that the pH of the water in the panels is between 6 and 8; otherwise corrosion will be a problem. The safest approach is to use distilled water as Thomason and other experts do, or to include a water softener in your system between the supply and your collector panels.

Piping

You will notice that the piping specified in Figure 19 is labeled PVC (Polyvinyl Chloride). This is a plastic pipe able to withstand high temperatures, up to the boiling point of water. Plastic pipe is used because of the possibility of corrosion if the aluminum solar collector panels are mixed with pipes made of a different metal. The amateur plumber will also find that plastic pipe is much easier to work with. Joints are cemented instead of soldered. The supply house where you buy your material can supply you with full directions on working with PVC or CPVC (Commercial Polyvinyl Chloride), which is a better grade of plastic.

Manifolds

A distribution system for the working fluid flowing to the solar collector panels is necessary. (See Figure 19.) These manifolds are fabricated from ¾-inch CPVC pipe and fittings. The distance between the "tee" fittings is 3 feet if the panels are used vertically, 8 feet if they are used horizontally. Each tee consists of a ¾-inch to ½-inch reducing tee, a plastic adapter to go from a socket to a threaded fitting and a threaded fitting with a barb connector on one side to accept the flexible hose from the Sol-R-Tech panel. It should be noted that this connection from panel to manifold could be a plastic coupler or, if using copper pipes, a brazed solid copper tube. The manifold pipe must be no smaller than the ¾ inch specified, but larger sizes will not affect the system's performance adversely. Always connect the feeder line to the manifold next to the first panel. Connect the outlet line from the manifold opposite the last panel in the series. This will equalize the flow rate and pressure in all panels.

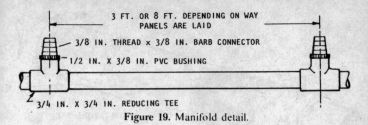

Figure 19. Manifold detail.

Panel Connectors

Sol-R-Tech solar collector panels are shipped with ⅜-inch NPT (National Pipe Thread) threaded connector blocks in place. A special plastic right angle fitting is included, to be installed in the connector block before the panels are put on the roof. (See Figure 20.) Use Teflon tape or pipe thread compound when screwing this fitting into place. A piece of ½-inch O.D. flexible plastic tubing is then forced onto the fitting and held in place with a stainless steel hose clamp.

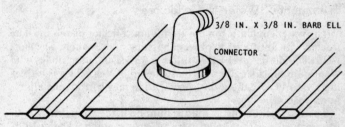

Figure 20. Panel connector detail.

Getter Column

The purpose of the getter column is to prevent corrosion in the solar collector panels. A piece of aluminum screen is used to attract the corrosive elements from the water before it reaches the aluminum panels. (See Figure 21.)

Getter columns can be made from PVC drum traps or from radiator hose. In order to use radiator hose, the ¾-inch CPVC feed line must be increased to 1½ inches with the proper adapters. Put a piece of rolled-up aluminum screen inside the hose. Get as much screen in as you can without completely blocking the flow of water through the hose. Splice the radiator hose into the feed line in

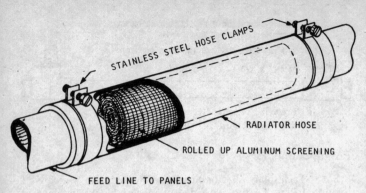

Figure 21. Getter Column.

place of the section of pipe you cut out, using stainless steel hose clamps.

Be absolutely sure that the screen you use is aluminum. Coatings or lacquers will defeat the purpose of the column, while using another metal will promote corrosion instead of stopping it.

Treatment of Water in Collectors

The water circulating in the aluminum collector panels must be treated with antifreeze to prevent bursting of the panels. The antifreeze should be specified as suitable for use in automobiles with aluminum block engines, such as Chevrolet Vega. This is important; antifreeze without this specification may corrode the aluminum panels.

The final heat transfer fluid should be made from 1 part antifreeze and 1 part distilled water (50% v/v solution). The distilled water can be purchased in local drugstores.

How to Attach a Liquid-Type System to the Roof

The following procedure is applicable to either new or existing houses. If a new house, a layer of waterproof roofing, such as asphalt roll-roofing, should be put on the roof sheathing first. Only a few inches of overlap is necessary, rather than the 100% overlap normally used with this material. It is important that the roof under the solar collector panels be waterproof, although it does not have to be designed to stand up to exposure to weather. On an existing house, the panels can be laid directly over the present roofing.

You will need a set of carpenter's roof staging brackets in order to work safely on the roof while installing the panels. These can usually be obtained at a local hardware rental store.

1. The first step is to lay out the complete collector array on the roof, using carpenter's chalk line. Be sure that the layout you made on paper is going to work out in practice. (See Figure 22.)

2. Decide how you will install the collector manifold (the header pipe with outlets coming off it to be connected to the panels). Will it be built along the top of the roof (see Figure 23) or will each panel be connected to an interior manifold running within the attic space? Each system has advantages. The first allows for fewer holes that need to be sealed and caulked. The second is easier to fabricate. The following steps describe the roof manifold construction in detail but also apply in principle to collectors plumbed individually through the roof.

3. Build the long wooden box which will house the top-collector manifold as shown in Figure 23. The length of this box depends, of course, on the length of your panel array and, accordingly, your header pipe. The outer width is about 10 inches, allowing at least 2 inches inside for the header. The wood beams themselves can be 1 inch by 4 inches, and the upper piece should be cut at an angle as shown in Figure 28b, for convenience later in sealing with flashing (described in step 9). The bottom wood piece should have holes provided for the ½-inch manifold pipes to the panels. The top of the box should be left off until the panels and manifold are in place.

4. Add the left side wooden trim piece, shown in Figure 24, down the entire left side of the collector array. It can be 1 inch by 4 inches thick, extending the height of your panels (usually 8 feet) and attached to the manifold box.

5. The top row of panels can now be laid in place (see Figure 25). Start at the left side and work to the right, simply laying the panels loosely on the roof. A strip of wood can be tacked in place along the bottom of the panels to keep them from sliding down the roof.

6. The top row of panels can now be fastened in place using aluminum T-sections and gasketed wood screws. These T-sections (see cross-section in Figure 26 blow-up) go between each panel, extending for the length of the panels, and thereby separating them slightly. They are fastened with wood screws to the wooden frame of each panel.

7. Add the right-hand wooden trim piece down the entire right side of the collector array. (See Figure 27.)

75

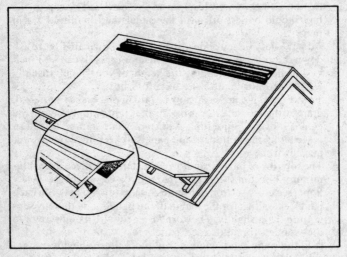

Figure 22. Chalk layout.

Figure 23. Top manifold box.

8. Put the top manifold in place and connect it to the panel with flexible ½-inch tubing or PVC pipes. The manifold can be assembled on the ground and lifted into place with the help of an assistant, or assembled from pre-cut pieces on the roof. (See Figures 28a and 28b.)

9. Nail the top of the manifold box and finish the flashing as shown in Figure 28b. The flashing is a sheet of aluminum or galvanized steel which covers the edge of the manifold box and the roof, thereby preventing any leaks in the crack between them. It can be affixed to the box and roof with nails, and further sealed with an adhesive or caulking suitable for your particular roof.

   If you live in a climate where the temperature often drops below zero during the winter, stuff the box with fiberglass insulation before you close it, to augment insulation of the manifold pipes.

10. Construct the bottom manifold box according to Figure 29 in the same manner as the top box except that the lower beam is cut on an angle and the upper beam contains the holes for the manifold pipes. Here again, leave the top off.

11. Put the bottom manifold in the box and connect it to the panels. (See Figures 30a and 30b.) Stuff the box with insulation if necessary, and put the top on the box. Complete the flashing as shown in Figure 29b.

12. Double check all wood screws to be sure the panels are being held evenly and firmly in place. Check the flashing for bubbles in the sealant, and to be certain it is adhering all along its length.

DESIGNING AN ADEQUATE TEMPERATURE CONTROL SYSTEM

Whether you decide on an air-type or liquid-type solar heater, it is important to have a suitable automatic control system. In the case of the air system, the thermostats activate and shut off your blowers and exhaust fans as needed, to keep your house comfortable. With a liquid system, similar controls turn the necessary pumps on and off as needed.

It may be possible to adapt your present thermostat controls, whether in a forced-air system or hydronic system, to do the job when you add a solar heater. Since there are so many makes of home temperature controls, our most helpful suggestion is to proceed as follows: First, find out the manufacturer and model number of your existing control system. Get the instruction book from

77

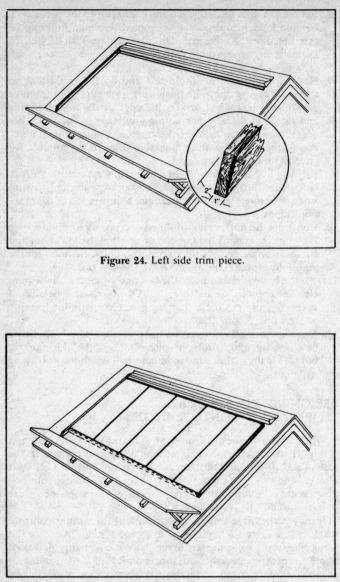

Figure 24. Left side trim piece.

Figure 25. Panels laid in place.

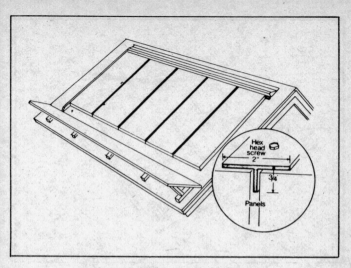

Figure 26. Panels fastened with T-sections.

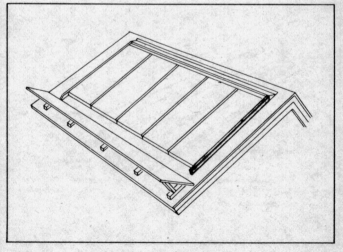

Figure 27. Right side trim piece.

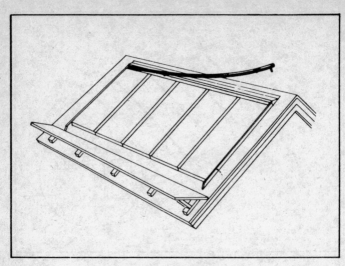

Figure 28a. Top manifold put in place.

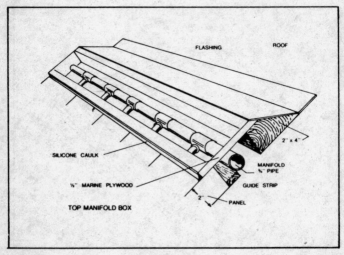

FLASHING ROOF

SILICONE CAULK

½" MARINE PLYWOOD

TOP MANIFOLD BOX

2" x 4"

MANIFOLD
¾" PIPE

GUIDE STRIP

2" PANEL

Figure 28b. Detail.

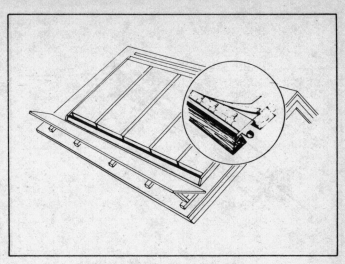

Figure 29a. Bottom manifold box.

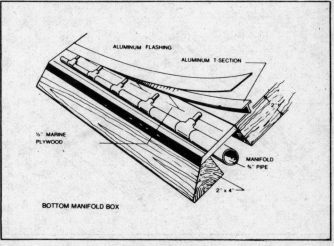

ALUMINUM FLASHING

ALUMINUM T-SECTION

½" MARINE
PLYWOOD

MANIFOLD
¾" PIPE

2" x 4"

BOTTOM MANIFOLD BOX

Figure 29b. Detail.

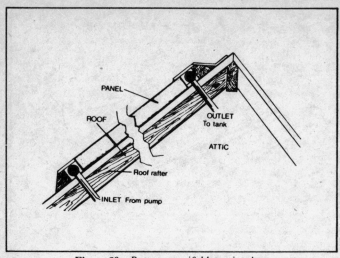

Figure 30a. Bottom manifold put in place.

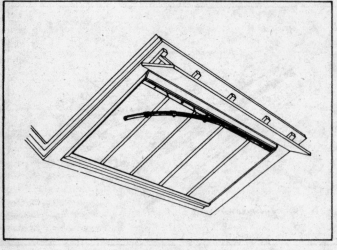

Figure 30b. Detail.

that manufacturer, or from a local distributor or dealer for this make, getting the name from the yellow pages. After you have this information, you may be able to make whatever modifications are required on your own. If not, consult a professional—an electrical contractor, licensed electrician, or heating and ventilating firm.

Your best solution, however, may be to use a combination of your present standard thermostatic control and a differential thermostat designed specifically for solar systems. It will be helpful for you to read the information in the next few pages and study diagrams describing two of the control systems now being used successfully for automatic operation of both air and liquid solar heaters. Either system can be used if your installation is in a new building, or if you are remodeling an older building to include a solar furnace.

Automatic Control for Air-Type Solar Furnaces

Figure 31 is a block diagram of the control system used in several air solar furnaces. This shows the Model 74-116 multi-process controller made by Zia Associates in Boulder, Colorado, together with thermistor temperature sensors and other hardware.

Whenever the standard thermostat, T4, in your home calls for heat, this solid-state controller makes a decision. If the temperature in the rock storage bin, measured by T2, is greater than a preset temperature, T3, at the controller, then the distribution fan is turned on to send hot air from the bin to the house.

You can adjust the level of T3 by any 10° increment from 60° to 160° F. For an air-type collector and rock storage such as the ISC system, T3 may be set as low as 90°, whereas for a liquid-type system such as Garden Way, Piper, Corning, Revere, Reynolds, PPG, Sunway, SAV, SolarSan, or any of the others producing truly hot water, T3 may be set at the upper end, 160°.

Meanwhile, the solar collector panel temperature, T1, is measured and compared with storage bin temperature, T2. (T2 could also measure a large hot water storage tank in a liquid system.) Whenever T1 is greater than T2, the collection fan (or pump) is activated and directs heat from the collector into storage. It's a differential system similar to that of Rho Sigma described later in this chapter.

As long as the temperature of the storage bin (or reservoir), T2, is greater than the setting of the thermostat, T3, in your controller, your rooms will be heated by solar energy. That's why the hot air system, preset for 90°, is quite efficient. However, when T2 is less than T3—the bin is colder than your home's heating needs—an

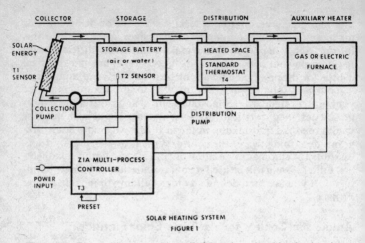

SOLAR HEATING SYSTEM
FIGURE 1

Figure 31. Zia multi-process controller for air or liquid solar heating.

auxiliary furnace is automatically started by the action of T4 and the controller.

Automatic Control for Liquid Solar Heaters

Another popular control system for solar heaters is made by Rho Sigma, Tarzana, California. The Rho Sigma differential thermostat, like the Zia unit just described, can be used with either liquid or air solar systems. In the diagram of Figure 32, explained in the next few pages, you'll learn how the Rho Sigma controller is applied to liquid-type solar heaters.

A basic reason why Robert J. Schlesinger, president of Rho Sigma, and others have developed differential thermostats for solar systems is that you can't control the temperature of water (or air) in a flat plate collector on your roof simply by turning a knob. In a conventional house you have two kinds of thermostats. One you set for room temperature, typically 68° to 70° F for heating. The other is for hot water at 140° F. When either temperature is below the set point, your heating system goes on automatically. If the temperature exceeds the setting, the heat goes off. Because the heat source in your usual system is much hotter than the desired temperature, there's no need to know the temperature at the burner flame. So each of the two conventional thermostats needs to sense only a single temperature—that of a room or of hot water.

But in a solar heating system that includes a pump, it is impor-

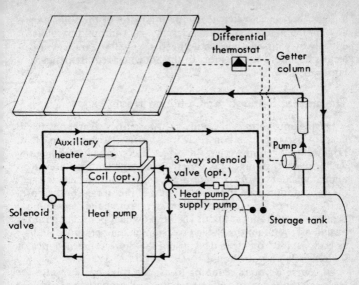

Figure 32. The Rho Sigma differential thermostat system.

tant to sense two temperatures and compare them: the temperature of the water in the solar collector panels, and the water in the storage tank. Hence the need for a different kind of thermostat, a differential device which measures both temperatures, compares them, and based on that comparison sends out an electrical control signal which turns the pump on or off. In brief, it is a thermostat with a brain.

During the day when the sun is shining, water in the solar panels will soon become hotter than the gallons in storage. So the pump operates to keep water recirculating through the system, building up heat energy in the storage tank for use in heating the house or domestic water. At night, or on heavily overcast days, the collector water temperature tends to fall below that of the storage tank. If your pump continued to operate then, it would defeat the purpose of the system by removing heat energy from storage.

So the automatic solution is to use a differential thermostat which senses and compares collector and storage tank temperatures, and thus controls the daily on-off cycling of the pump for most efficient operation of the solar heating system.

Let's apply this principle to the installation diagrammed in Figure 33. The 50-gallon storage tank is the reservoir for solar energy and serves as a preheating system ahead of a conventional 30-gallon gas or electric water heater. Using many of the solar panels now on the market—including designs you can build partially or entirely by yourself—you would need about 50 square feet of collectors to supply 60% or more of the hot water requirements for a family of four.

Thermostat Settings for Optimum Heat

During long periods of cloudy, rainy or snowy weather, you will need to use the 30-gallon fuel-fired heater. However, on clear, bright days, even in winter, the effect of continued solar heating will bring your storage tank water to very high temperatures. Rho Sigma president, Robert Schlesinger, suggests adding a "high set" thermostat immersed in the upper part of the storage tank to prevent transfer of too much scalding water through the conventional heater to your home at about 160° F, at which point the pump is turned off. Without this safeguard, storage temperature could go as high as 180° on clear days, using the more efficient types of solar panels on your house.

Of course if you're planning to use the solar-heated water for warming the rooms of your house, you will probably use a bigger storage tank. Even if you don't, a higher temperature will simply

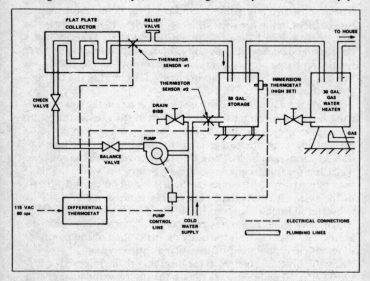

Figure 33. Typical solar hot water system using a differential thermostat.

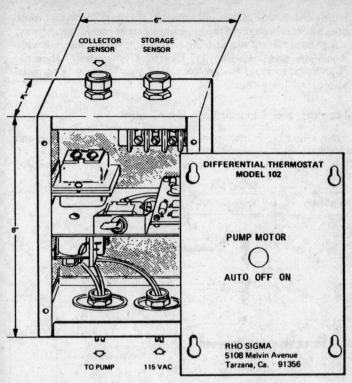

Figure 34. Cutaway view of differential thermostat.

mean a slight increase in space heating and no necessity for a high-set thermostat.

Indoor and Outdoor Thermostat Models

There are both indoor and outdoor models of differential thermostats. The Rho Sigma Model 102 for hot water systems is priced at $70, and the Model 104 for use with collectors containing antifreeze and using heat exchangers is $85. A cutaway view of Model 102 appears in Figure 34.

With either type of differential thermostat, there are two matched thermistors required as temperature sensors—in the solar collector exit, and at the bottom of the storage tank, or in the exit line leading to the pump. You can insert these thermistors into the piping system with tee fittings as shown in Figure 35. Wiring from the sensors is brought to the appropriate terminals in the thermostat control box. Make sure the unit you buy, as is the case with Rho Sigma and Deko-Labs TC-1, carries Underwriters Laboratory approval to avoid any problems with your insurance company or local building codes.

Manual and Automatic Pump Control

Note that the control panel of the differential thermostat includes a switch to permit manual as well as automatic control of the recirculating pump. This makes it easier for you to test the sys-

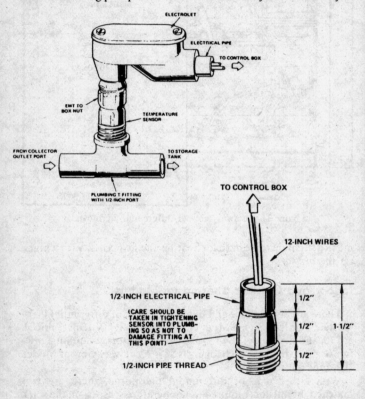

Figure 35. Matched sensors for use with differential thermostat.

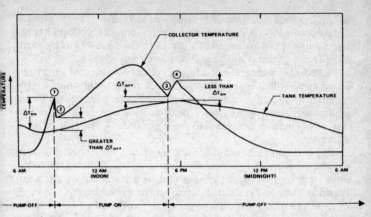

Figure 36. Pump cycling under control of the differential thermostat.

tem when you have installed your solar panels, storage tank, pump and the rest of the heating system.

In Figure 36 is a temperature versus time graph that shows how a differential thermostat will control your pump on a typical day, and why the differential settings and a hysteresis circuit prevent unnecessary cycling of the system. Typically the differential turn-on temperature is 20° plus or minus 3°, and the differential turn-off range is 3° plus or minus 1°.

Following the curve of collector temperature in Figure 36, the sun's heat on your panel array turns the pump on at point 1. Cooler water in the storage tank carries the heat away and causes an initial drop to point 2. However, this temperature difference is greater than 4°, the maximum differential between collector and tank for turn-off, and thus the pump keeps working.

Late in the day at point 3 on the collector curve, the temperature has dropped to the point where the pump is turned off. Now water in the panels is still being heated by the sun and not being carried off to storage, so there is a sharp rise in temperature. But this rise is less than 17°, the minimum differential for pump turn-on. Thus the pump stays off.

The Storage Tank is a Stable Heat Reservoir

You'll notice that, while there are marked variations in collector fluid temperatures during the 24 hours, the temperature in the storage tank shows far less variation. It's your heat reservoir for stability. This is why many larger homes equipped for solar space

89

heating as well as hot water heating contain tanks of 2,000 gallons or more, some in rock bins, to store large amounts of solar energy. In this way it is possible to continue to utilize solar heating even after five or six days of continuously cloudy weather.

Figure 32 is a diagram of a liquid system for space heating prepared by Garden Way Laboratories, Charlotte, Vermont, which includes all the elements discussed in previous pages as well as a heat pump, which serves as a type of auxiliary heater. This pump not only generates heat (usually from electricity), but distributes it as well.

This is the kind of system that is most easily installed in a new building, because of the space required to install such major elements as a large hot water storage tank and a heat pump. The heat pump is used to heat either liquid or air, whichever is the heat transfer medium, and then to pump the hot fluid or air to areas where it will do useful work—such as heating your house. You

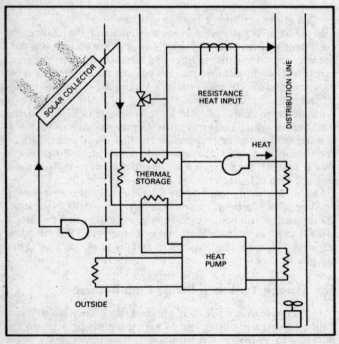

Figure 37. Solar heating with heat pump auxiliary.

might consider a heat pump as a combination of heat exchanger and input-output pump.

To see how a heat pump is used as an auxiliary heater with a forced-air heating system, look at Figure 37, a schematic drawing prepared by PPG Industries. Note that in this case air is pumped into the heat pump from the outside. The rock bin labeled Thermal Storage is, as we've discussed, used as a heat reservoir getting inputs from the solar collector, heat pump and a possible auxiliary heater noted as Resistance Heat Input. The forced-air heating system gets inputs as required from the rock bin, heat pump and auxiliary heater.

PPG does not show a differential thermostat in this schematic, but you can readily see how it fits in by referring back to Figure 33.

Sample Buildings Using Liquid-Type Solar Heaters

New Hampshire House's Solar Heating System

A house in Etna, New Hampshire, with two stories and 2,100 square feet of heated area, has a solar heater designed by John DeVries, vice president of Sol-R-Tech and designer of the Sol-R-Tech collector panels. On the 45° roof areas of this house (30° on a dormer area) are twenty-five of these panels described in Chapter 3. Outer glazing is Alsunite fiberglass-reinforced polyester, and the heat transfer fluid is silicone oil.

There is a 4,000-gallon water storage tank of poured concrete and an auxiliary 500-gallon tank of precast concrete, with both tanks beneath the floor slab of the house. Two heat pumps are used to extract energy from the tank water and deliver it to the rooms by forced hot air. These heat pumps are York Triton pumps, DW-30H and DW-40H.

While there are auxiliary electric air heaters in the ducts, this solar heater handles 100% of the heating of a large house in cold New Hampshire—near Hanover, site of the famous Dartmouth Winter Carnival. Domestic hot water is preheated by the solar system. There is a carrythrough of two days when the weather is too bleak for sunlight.

Solar Heating for a Store

Shown in Figure 38 is the Custom Leather Boutique Building, White River Junction, Vermont, with one and a half stories and 2,100 square feet heated by a similar liquid system designed by Sol-R-Tech's DeVries. In this case, there are 910 square feet of

Figure 38. Even in winter, array of Sol-R-Tech panels supplies Vermont store with over 90% of room heating and hot water.

Sol-R-Tech collectors on a 45° roof. The storage tank under a concrete slab is made of concrete with a waterproof epoxy lining, and holds 5,000 gallons of water. There are three York Triton heat pumps to take heat from the big water tank, so long as the water is above 45° F, and deliver the energy through a forced-air system.

In this case, antifreeze is not used. Collectors are drained into the storage tank at night when there's a chance of freezing weather.

Telephone Company Building Solar Heated

Since American Telephone and Telegraph (AT&T) and its Bell Laboratories have pioneered many inventions, largely in electronics, it's not surprising that a major AT&T subsidiary, New England Telephone and Telegraph, has a building in North Chelsford, Massachusetts, which is using solar heating.

A 3,800-square-foot building thirty miles northwest of Boston has 400 square feet of collector panels in two arrays, each with ten panels. Half of the collectors are made by Revere Copper and Brass, the others by Daylin. There are specifications on both types in Chapter 3.

Fluid in the collectors is an antifreeze mixture, and on a bright sunny day its temperature will reach 200° by midafternoon. There is a single cover of glass on both types of panels.

Water storage is in a well-insulated steel tank holding 500 gallons and situated outside the telephone building, because this solar system is a retrofit, installed for testing on an existing structure.

About 50% of the building's heat load is delivered by this solar system, using a fan-coil method for delivering heat stored in the water tank. (See the Piper discussion at the end of this chapter for more details.) Auxiliary heat is provided by an oil-fired hot air furnace. For the next installation, the telephone company will probably use more solar collectors, a larger storage tank, and perhaps heat pumps so that the solar system will handle a far larger percentage of the heating load.

Award-Winning Solar House in Connecticut

A three-bedroom beachfront home in Connecticut derives more than 60% of its space heating and hot water from liquid-type solar panels made by Sunworks. Designed by Donald Watson, AIA, with technical assistance from Everett Barber, it has won several architectural awards (see Figure 39).

Figure 40 shows a cross-section of this house. From three arrays of collector panels, the sun-heated water moves into a large 2,000-gallon storage tank. This heat bank preserves enough heat for two or three sunless days in winter.

From the cylindrical tank shown in the diagram, the hot water is piped through the house to fan-coil heat exchangers. When the home thermostats call for heating the rooms, air is blown over copper coils containing sun-heated water. Some of this hot water also is delivered through copper pipes to bathrooms, kitchen, and laundry.

If there is a prolonged stormy period, a conventional oil-fired water heater is turned on automatically to carry the load until the sun's heat is again doing the job.

Colorado Test House with Liquid Solar Heaters

In two test houses in Fort Collins, Colorado, built by Colorado State University, the solar engineering was under the direction of Dr. George O. G. Löf and D. S. Ward. One house has 710 square feet of PPG collector panels, having Olin Roll-Bond aluminum sheet with extruded channels as the absorber plates, painted non-selective black. This system uses an antifreeze mixture in the panels, a heat exchanger, and a heavily insulated steel tank holding 1,100 gallons of water in the basement.

Figure 39. Beachfront home with liquid-type panels by Sunworks has won several architectural awards.

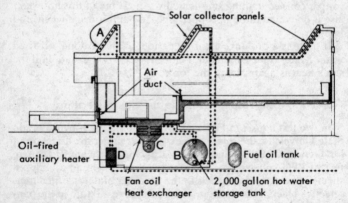

Figure 40. Cross-section of prize-winning solar house using a liquid heat transfer system.

Referring back to the beginning of this chapter, Table 1 on collection efficiency versus ambient temperature should be considered as somewhat conservative. Measurements by Dr. Löf and associates in the Colorado house show a collection efficiency of 45% while the temperature difference (T_d) between outlet water from the collector and the outside air is 100° F and the irradiation or insolation level is 300 Btu per square foot per hour.

This 3,000-square-foot house gets 75% of its space heating and preheating of domestic hot water from solar energy. Carrythrough is two days.

A second Colorado State house uses Corning Glass collectors described and illustrated in Chapter 3. This type of collector has considerably greater efficiency as follows:

Collector Efficiency (%)	T_d(°F)	T_{liquid} (°F)	T_{air} (°F)
70%	150°	200°	50°
20%	300°	350°	50°

Note that this test data is much more impressive than that tabulated earlier in this chapter, but the end result is similar. Efficiency of the collector *decreases* as the difference between collector fluid (or air) temperature and the outside air temperature *increases*.

In this house, an antifreeze mixture is again the material used in the collectors, and the storage tank and other parts of the system are nearly identical. Preliminary results indicate some gains over the other system because of the great heat collecting efficiency of the Corning evacuated glass tubular collector system.

A third test house at Fort Collins, Colorado, designed by the same team, uses a Solaron air collector system and a rock bin. Its efficiency appears to be quite comparable to that of the first house described.

Efficient Liquid-Type System for Commercial Buildings and Homes

About seven years ago, a building contractor named James R. Piper began building apartment houses using a system for space and water heating which he called a hydronic series loop. In basic terms this consisted of running one line of copper pipe 1½ inches in diameter all the way through a building containing two hundred units. This line carried hot water, and the heat was replenished at fifteen boiler stations placed in such a way as to even out the heating load between sections of the line. He began by using conventional gas-fired boilers at relatively low temperatures—typically water never heated above 143°F.

This hot water pipe was looped into a heat exchanger in each apartment, and air was blown over the hot pipe when the apartment needed heating, as called for by a thermostat. Domestic hot water for each apartment was tapped from the same series loop of copper pipe. At each boiler station there was a make-up supply of cold water which was fed in on demand any time the water in a section of the loop between boilers—each handling about thirteen apartments—needed replenishing.

After considerable experience in building large apartment complexes in various parts of California, Piper was able to prove that his low-temperature series loop for handling hot water and space heating was highly efficient. Bills for consumption of natural gas were far lower than in similar large apartment buildings.

At the same time, Piper became interested in solar heating as a supplementary source of energy. It was easy for him to see that solar panels on the roof of an apartment house would heat water in a storage tank and thus preheat water entering his boiler stations, which burned gas only when necessary. In 1969, when he first proposed doing this, Piper says: "I couldn't find a single owner of an apartment complex who was interested in solar energy. Even though I could show them that it would add relatively little to the total cost of the building. And it would pay for itself in a few years. The reason was that the owners don't pay the gas bills, the tenants do."

Now the picture has changed drastically because of public interest in solar energy and fears of even higher gas prices. In the plant of Piper Hydro in Anaheim, California, solar panels are being produced for many types of buildings, including homes, apartment houses and commercial buildings. The Piper solar panel is 8 feet long by 2 feet wide and utilizes ½-inch blackened copper tubes, nine per panel, attached to a corrugated galvanized steel sheet—common roofing material. There are 1½-inch copper headers at each end which mate with the standard pipe in the hydronic loop. Covering the panel is a sheet of 4-mil Tedlar, while under and around the metal collector is 3-inch-thick fiberglass insulation. The entire panel is enclosed in a galvanized steel box much like the do-it-yourself panel described in Chapter 3.

When building a new, solar-heated house, the additional cost of a hydronic series loop is about $2,000 for a house with 2,500 square feet.

Installations of the Piper Hydro system include a school in San Diego County; four homes in Spokane, Washington, one with solar panels for water heating only, the others with the complete system; the home of a builder in San Jose; homes of architects in Palos Verdes and Newport Beach, California. Another solar system recently installed in a home in Mt. Airy, Maryland, really proves the point: "Their utility bill for December and January was less that $50, while a house the same size nearby cost $250 a month for the same cold period," Piper says.

UPDATED INFORMATION

Since the first edition was published, Sol-R-Tech has developed a packaged liquid solar heating panel which is offered at a price of less than $200. Some specifications are:

Center to center of frame 96" x 32"
Outside dimensions (model 3296-1) 99⅝" x 35⅝" x 3⅝"
  (not including manifolds)
Net weight 2.3 lbs. to 2.8 lbs. per square foot
Gross collector area 21.33 sq. ft.
Net collector area 18.73 sq. ft.
Connections ½" nominal copper fittings
Recommended flow rate 0.33 g.p.m.
Pressure drop at 0.33 g.p.b. is 0.53 psi
Windload test 15 minute 65 m.p.h. test shows nondetrimental deflection in outer glazing. A retaining bar can be added to units used in very high winds.

5 / Building A Solar Water Heater

One of the least expensive applications of solar energy is the heating of domestic water. Since we previously discussed solar heating systems that included water heaters, we will now focus on those systems designed specifically for heating domestic water.

THE ECONOMICAL HITACHI HI-HEATER

With the growing competition between solar manufacturers, and a growing solar market, the cost of solar water heaters is coming down. Already, you might be surprised at the low cost of a simple solar system for heating water in your home. For less than $400, not counting delivery and installation, you can get a single-panel Hitachi Hi-Heater which has a capacity of 44 gallons of hot water! A product of Hitachi Chemical Company, this solar water heater has been in mass production for several years. According to the manufacturer, tens of thousands of these heaters are in use in Japan and other Asian countries.

An installation of a single Hitachi panel (with attached water tank above it) is shown in Figure 41. The panel comes already assembled. Construction of this panel is economical and utilitarian. There are six heat-collecting cylinders made of black, rigid polyethylene plastic mounted in a steel box containing about ⅓ inch of styrofoam insulation under the cylinders. Dimensions of the outer box are 3 feet 9 inches by 6 feet in area and about 8½ inches deep.

Over the box, covering the black cylinders as a heat trap, is a sheet of transparent polycarbonate plastic. There is also a thin layer of aluminum foil over the layer of styrofoam insulation. This foil reflects solar heat from the sides and bottom of the steel box and tends to direct it at the black plastic cylinders which contain water.

Figure 41. Installation of a Hitachi solar water heater.

Mounting the Panel

For best operation, this panel should be mounted at an angle of at least 30° from the horizontal. An inclination of 15° is necessary to permit thermosiphoning, the upward movement of the water in the solar cylinders as a result of heating. Since a panel weighs about 540 pounds when full of water, it is well to check your roof before installing it. However, considering the size of the panel, this weight is only 25 pounds per square foot, so any well-constructed roof should support one or more panels.

Typical methods of installing the heater are shown in Figure 42 for flat and pitched roofs. The same rack used on a flat roof can also be used on the ground, in your backyard. (See Chapter 8 for details on rack construction.) To install the Hitachi heater, you drive coach screws into the roof or rack and attach stainless steel wires to the screws—this hardware being part of the kit with the solar heater. Then you fasten the wires to the hooks at the four corners of the heater box, leaving no slack in the wires. The transparent plastic cover sheet is secured to the steel box by means of stainless steel bands and retainer clips, which are also provided.

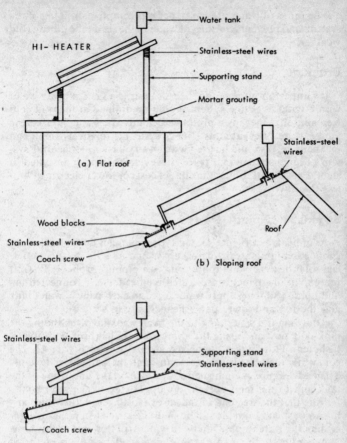

Figure 42. Installation methods for putting Hitachi panels on roofs with various slopes.

Use of a Ball Tap

A ball tap is installed at the inlet side of the water tank. It is to provide automatic control of the water supply to the solar cylinders so that as hot water is drawn from the system, it is replaced by more water and the water tank stays full. This ball tap will handle a wide range of water pressures, permits complete draining of the system, and prevents freezing of water in the feed pipe during cold weather.

Drainage

It's important to note that because this is a system using water with no additives, the system should be drained at night in cold weather, or if there is a prolonged cloudy spell and subfreezing temperatures. Because this Hitachi system is relatively small and inexpensive, it should be used with an auxiliary water heating system of any conventional type. However, if properly installed, it should pay for itself in reduced fuel costs or lower electric bills in less than two years.

Suggested Piping

As indicated in Figure 43, this type of panel is designed for easy connection into an existing water system. As with most commercial solar heaters, you supply your own piping—probably PVC. If more than one panel is used, then the headers are connected top and bottom so that supply water is fed to each panel's water tank and the top and lower panel headers are in series.

Attaching the water tank to the heater box involves sliding the tank on a post and using the appropriate bolts (provided) for attachment. Plumbing is connected as indicated in Figure 43. The cold water feed line is brought in through the ball tap to the tank. Then a feed pipe is connected to a nozzle on the underside of the tank and to the inlet port on the upper side of the steel box.

Note that the free end of the air exhaust hose, which vents any hot air or steam from the system, must be raised to a position higher than the water tank. Otherwise water will flow out of this hose.

According to the manufacturer, polyethylene pipe is preferred because it's cheap, relatively easy to work, and not apt to crack in freezing weather. Other alternatives are PVC, copper or galvanized pipe. If your experience as a plumber is limited, by all means get advice from a neighboring plumbing supply store. The owner may be able to steer you toward an expert whose rates are reasonable for the tough jobs, in case you don't have a friend with plumbing expertise. In any case, make sure all your connections and joints are water tight.

With this kind of solar heater, if the pipe supplying water from the panel(s) is leaking, your entire system may fail to operate. This is because, while the hot water supply pipe is leaking, cold water is

102

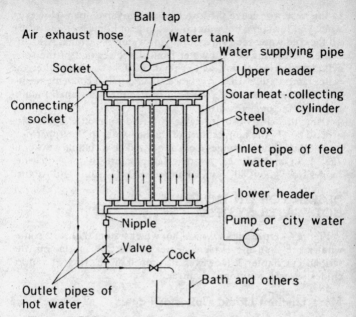

Figure 43. Piping connections for Hitachi solar collector panel.

automatically fed through the tank to the collector cylinders. It is likely that too much cold water will enter the cylinders to permit adequate heating. So be careful to avoid leaks.

When there is adequate water pressure and it never falls below a level where your supply line will keep the solar water tank full, it's desirable to draw your hot water from the upper header of the heating panel. You can easily tell whether your water pressure is adequate by opening the valve from your cold water supply, closing the other valves, and observing whether the water tank fills. If it does and your pressure is good, the normal mode of operation is to keep the valve shown below the lower header closed except when you want to drain the system.

Need for Pumps

If the water pressure is too low, you may have to use a small pump to take the water up to the solar heater tank(s) on your roof. Should the water pressure continue to be low or variable, you may have to draw your hot water from the lower header of the heater.

103

In this case, you leave the lower hot water valve open. However, when you turn on a hot water faucet in your home, you must turn off the cold water feed valve supplying water to the tank. Also, because you're drawing hot water from the lower end of the solar cylinders, the initial flow will be relatively lukewarm and then the water temperature will rise. Hence, as a matter of convenience it's important to make sure your feed line to the roof tank(s) maintains satisfactory pressure.

Using this kind of solar heater with feed water at temperatures around 60° F, the temperature of the hot water in the upper portion of the cylinders ranges from about 100° F in January to about 155° F in midsummer. Average temperature of the hot water ranges from about 80° F in January to 120° F in July and August.

DESIGN FOR AN IMPROVED SOLAR HOT WATER SYSTEM

Figure 44 represents a typical hot water system that costs somewhat more than the Hitachi heater or the Sol-Therm unit described in Chapter 3. However, it is much more efficient and includes several additional features.

More Efficient Liquid Collector Panels

If you review the material in Chapter 3 on various types of solar collectors of the liquid type, you'll find many of the newer makes, including Corning, Garden Way, PPG, Revere, Reynolds, Sunsource (Daylin), Sunwater, Sunworks and others using metal absorber plates, heat water to higher temperatures than are achieved with plastic panels.

These collectors, including Edmondson's SolarSan unit described in Chapter 3, will often produce superheated water. In one of Edmondson's tests, the water from his collector outlet reached 235° F in January. This means that your heat exchanger in Figure 44 will rapidly heat water in the large solar storage tank.

Use of a Heat Exchanger

A second improvement is the use of a heat exchanger, which permits you to use an antifreeze mixture in your solar collectors in areas where freezing might occur. If your panels are made of aluminum, most manufacturers, like Garden Way, recommend using antifreeze and a corrosion inhibitor. Even if you live in a warm area, you should use either distilled water or include an anti-rust solution.

104

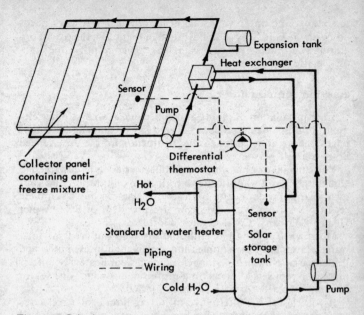

Figure 44. Solar hot water system designed by Garden Way Laboratories using antifreeze mixture in collectors.

Heat exchangers have previously been described in Chapter 3. A liquid-to-water exchanger can readily be made by a do-it-yourself person. When your hot outlet pipe from the solar panel array reaches the hot water tank, run the pipe into a serpentine consisting of many coils of copper or aluminum pipe. You might want to enclose these coils in a box lined with copper or aluminum sheet, as shown in Figure 44. This exchanger will heat the water in your storage tank as it continues to circulate past the solar-heated coils or their hot metal container.

Heat Storage and Temperature Control

Your heavily insulated storage tank can be as large as your family's needs for hot water. Typically a 50-gallon tank will be adequate. Some of the makers of commercial tanks are listed in Chapter 7.

Note that this system includes a differential thermostat. After reading Chapter 4, you're well aware of the advantages of installing such a differential temperature control system since it operates the pump or pumps in your solar heating system only when needed. With the kind of system illustrated in Figure 44, the cost of operating your pumps will be less than $1.50 per month in the highest electric rate areas.

Ask the Manufacturer

Since installation of piping and collectors, as well as other elements of this system, has been covered in Chapter 4 and previous pages, there's no need for further instruction for the experienced do-it-yourself person.

You can make or buy your solar collector panels as described in Chapter 3. If you want further information on how to build your own solar hot water heater from scratch, you can buy a complete set of plans from W. V. Morrow, Jr., president of Solar Water Heater Company, Coral Gables, Florida.

Morrow's company has built and installed hundreds of solar water heaters in Florida homes and apartment houses, motels and trailer parks. Now he's no longer building solar heaters but offers three different sets of plans for a professional system: one set for $35, any two sets for $50, all three for $60.

If you choose the alternative of buying liquid solar collectors from any of the manufacturers listed in Appendix 1, you will find them extremely helpful in supplying information as well as hardware.

For example, Garden Way Laboratories has available, at very reasonable cost—free, if you buy their hardware—a detailed instruction manual telling you how to install the Sol-R-Tech collector panels. Included is a list of other kinds of hardware, recommending everything from black paint to piping to heat exchangers and differential thermostats, all from approved sources.

Cost of Garden Way Solar Water Heater

As indicated in the description in Chapter 3, cost of the Sol-R-Tech collector is $7 per square foot, which appears to be competitive with most other designs. From the standpoint of architects, builders and handy persons, this kind of efficient panel, along with the plans and other materials for a complete solar heating system available from the manufacturer, is well worth consideration.

Table 2, supplied by Sol-R-Tech's Garden Way Laboratories, shows how many panels are needed for varying sizes of families in

106

Boston; Lincoln, Nebraska; San Antonio, Texas; and Hatteras, North Carolina.

Please note that both the Boston, Massachusetts as well as the Hatteras, North Carolina family could achieve 90-100% of their annual hot water demand by adding more panels. However, each has chosen to make a smaller initial investment while still reducing their hot water fuel consumption by 45-50%. Panels can always be added later.

If you consider the capital investment each household must make in light of today's escalating electricity and fuel rates, you quickly discover that each system pays for itself in four to eight years. If you also consider the measure of independence that solar hot water imparts, then the system becomes even more worthwhile.

Location	Boston, Massa-chusetts	Lincoln, Nebraska	San Antonio, Texas	Hatteras, North Carolina
Number of people in family	3	2	4	3
Percentage of heated solar water supplied	45-50%	90-100%	90-100%	45-50%
Costs				
Sol-R-Tech Solar Panels (#)	$504(3)	$672(4)	$840(5)	$336(2)
1,120 gal. tank	110	110	110	110
2 circulator pumps	140	140	140	140
Heat exchanger	100	100	100	100
Solar thermostat	85	85	85	85
Assorted plumbing	130	130	130	130
TOTAL	$1,069	$1,237	$1,405	$901

Table 2. Cost of a Sol-R-Tech solar hot water system based on number of people in household and location.

6 / Cooling Your House With Solar Heat

While it's natural to think of heating your house and providing domestic hot water with solar energy, it might seem strange to think of cooling a room with the sun's heat. But not so if you remember the principle on which a gas refrigerator works. The heat of a small gas flame turns a liquid refrigerant like freon into gas. This vapor then expands through cooling coils; but as the gas travels away from the heat of the flame, it returns to its liquid state. In doing so, the gas gives up its heat and thereby becomes cold. The heat is exhausted from the refrigerator, and the cold liquid then proceeds to chill its surroundings.

METHODS OF SOLAR COOLING

There are several ways in which solar houses are cooled in the summer, and this chapter will review them all. Probably the most popular is to use solar heat for so-called absorption cooling, applying the sun's heat instead of the gas flame in a refrigeration system.

Another popular technique utilizes a heat pump or heat exchanger. For example, you might use water-type solar panels on the roof of your house as a source of cooling at night. You pump water from a storage tank to the roof, assuming you live where the nights are cool, and return the water chilled by the night air to storage. Then, during the day, your heat pump removes hot air from the rooms, passes this air by the cold water tank, and returns the cooled air to the rooms.

A similar approach proved practical is using air as the medium of heat exchange. Your fan system sucks in air at night when it is cool and uses it to chill a storage medium. One favorite method mentioned earlier is to provide a large bin filled with egg-sized stones. During the night this bin becomes thoroughly chilled. Then, during the day, heated air from the house is passed over the rock bin, which is usually underground or in a basement. Air cooled by the rocks is then blown back into the house to keep its temperature comfortable.

The Skytherm Southwest House, previously mentioned, utilizes giant water beds on a flat roof, with sliding insulated panels over the plastic bags containing water. In a hot, dry climate such as Arizona, Nevada, and parts of California, this kind of temperature control works well. In the summer, at night, the water-filled bags are exposed to the sky and lose energy to it by convection and radiation. Thus the water becomes cool and chills the metal ceilings of the rooms below the plastic bags. During the day, with insulated covers automatically moved into place to shield the water bags from sunlight, the bags gain very little solar energy. However, they do absorb heat from the ceilings as the rooms begin to warm up, and thus they keep the house at a temperature of about 75° when the outside ambient is over 90° F.

The Importance of Insulation

One of the most important considerations stressed by experienced architects and builders, in cooling as well as heating, is to be sure your building is well insulated. Also if you're planning new construction, see to it that you provide adequate overhangs or awnings for windows facing south. In fact, since roof areas facing south are the ideal place to put solar panels, consider making this portion of your roof from solar collectors. There are many suitable types of panels described in Chapter 3. Then install the panels in such a way as to provide shading of the south-facing windows from the sun. This will assure you of an efficient solar collector and a cooler house.

Another procedure that's being followed in many new commercial buildings, as well as homes, is to apply a reflective plastic film to the inside of your windows facing south. For example, a material called Solar-X is manufactured by Solar Control Products Corporation, Newton, Massachusetts. This film comes in rolls with a pressure-sensitive adhesive on the back so that it will stick to window glass. There are fully reflective and partially reflective films. For commercial installations it's possible to buy the film for application with a liquid adhesive that is sprayed on the glass.

The advantage of this reflective film system for windows and glass doors is that it rejects up to 67% of solar heat striking these areas on hot summer days, and thus reduces the amount of energy required for cooling the building. Also this reflective film saves fuel energy in the winter because it reduces the amount of radiated room heat being absorbed by the glass and then being conducted and radiated outside.

ABSORPTION COOLING IN LIQUID SYSTEMS

Ammonia System

As long ago as August 1961 at the United Nations Conference on New Sources of Energy in Rome, there were ten technical papers presented on the use of solar energy for cooling. In a discussion by N. G. Ashar and A. R. Reti, both research assistants associated with the M.I.T. program of solar homes, the solar air-conditioning and heating system shown in Figure 45 was described.

From the solar collector panels on the roof, the water in the storage tank is raised to a temperature of about 120° F. At this point valves V_1, V_2, V_3 and V_7 are closed (automated by a thermostat not shown here), and sun-heated water from the collector goes directly through the coils of the vapor generator and is pumped back to the collectors on the roof. In this way, as the day goes along, the water in the panels gets hotter and brings the generator temperature up to an efficient level of about 140° F.

In the generator is a solution of 40% ammonia in water. As the solution is heated, it becomes a gaseous mixture of ammonia and water vapor. The ammonia gas is separated from the steam in the rectifier and then it is liquefied in a water-cooled condenser. Now the pure ammonia liquid passes through heat exchanger No. II and expands in the evaporator, where it becomes a gas. This latter process results in rapid cooling within the evaporator so that the coils of gas brought out from this cooling chamber are chilled. In the summer, as indicated in the upper right corner of Figure 45, air blown over these cold coils is then used to cool the house.

To complete the cycle, the cooled ammonia vapor is returned from the evaporator to the absorber where it's used to enrich a water solution which has come from the generator through heat exchanger No. I. The enriched mixture of ammonia and water is pumped back into the generator, to be heated again by sun-heated water from the solar collector panels on the roof.

As indicated with this kind of system, if you want to heat only the water in the storage tank, you close valves V_3, V_4, and V_7 and open V_1 and V_2. During the winter you can use the system for space heating by opening V_3, V_5 and V_9, while you close V_4, V_6 and V_8.

111

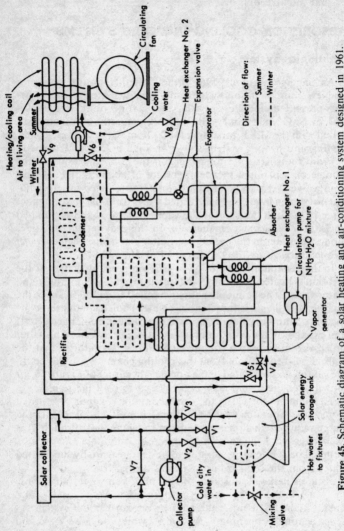

Figure 45. Schematic diagram of a solar heating and air-conditioning system designed in 1961.

Lithium Bromide System

In many parts of the world where there is ample sunshine, houses have been built with absorption cooling systems utilizing solar energy. In the Brisbane Solar House in Australia a lithium bromide absorber is used effectively, much like the ammonia system just described. This house has been cooled by solar heat since 1966. It has 1,318 square feet of air-conditioned living space. The solar collector there is a copper plate, 16 feet by 4 feet, with 1½-inch copper tubes soldered to it at 6-inch intervals. Both plate and tubing are painted with a selective black coating. There are two layers of glass above the copper plate to trap reflected infrared radiation and 3 inches of polyurethane foam insulation under the collector.

For storage of the sun-heated water, there is a 70-gallon tank. The medium used in the air-conditioning cycle is lithium bromide in aqueous solution, passing from generator through condenser and evaporator to absorber. Air is blown over chilled coils through ducts to cool the house. This building, sponsored by the University of Queensland, uses 3-inch plywood-faced polystyrene foam panels for walls and ceilings, with brick end walls and a floor of 6-inch concrete slab. Windows have double glass insulation. The roof is designed to shade a patio and north-facing windows, because it's in the Southern Hemisphere, and the roof is pitched at a 10° angle from the horizontal, facing north where the solar collector is mounted.

On clear days, when the outside temperature is well over 100°, this solar cooling system keeps indoor temperatures down to 70° F. The collector efficiency averages about 32% and the efficiency of the lithium bromide refrigeration system is 65%, so that the total efficiency is about 20%. One improvement made since the house was first built is the provision of an underground rock pile with 30 tons of fist-sized stones. Cooled air from the solar air-conditioning system is used to chill this rock reservoir, which acts as a storage tank for cold and makes the entire system more efficient.

Examples of Absorption Cooling

University of Florida Solar House

Another successful academic project is the University of Florida Solar House in Gainesville. This house has been in operation since 1963. On the roof are now 500 square feet of water-type collector panels consisting of aluminum sheets 10½ feet long by 4 feet wide with rectangular aluminum tubes clamped on the sheets at 4-to 6-inch intervals. Coating is a non-selective flat black paint and there is a single glazing of double-strength glass panes, each 4 feet by 1 foot. The water storage system is a 3,000-gallon lined steel tank insulated with 4 inches of styrofoam and placed above ground so that visitors can inspect it.

This hot water serves in the summer to provide 100% of the cooling of the building by operating an ammonia-water absorption system, which functions intermittently, as required. In winter, the solar system heats the house by circulating hot water from the storage tank through radiators. The same system provides all the domestic hot water needs and heats a swimming pool when necessary. There is no auxiliary energy source for cooling or heating in this installation. Solar panels face south and are sloped 30° from the horizontal.

NASA Solar Test Building

In Huntsville, Alabama, the solar test house built by the NASA Marshall Space Flight Center actually consists of three large office-type trailers joined together. Collector panels of aluminum with integral channels are coated with a highly selective black finish applied by electroplating. Glazing is 4-mil Tedlar supported by a steel mesh screen, and underneath the collector plates is 6 inches of dense fiberglass insulation.

Solar-heated water is stored in a horizontal 4,700-gallon aluminum tank containing deionized water—to prevent corrosion of the aluminum in the tank and panels—and having 3 feet of high-density fiberglass insulation. This NASA collector system produces water at temperatures up to 210° F, which supplies energy to a lithium bromide water absorption system made by Arkla Engineering Company. The system is calculated to have an efficiency rate of 67% and to provide, in standard air-conditioning terms, 3 tons of cooling for the three trailers under one specially built roof, facing south at a 45° slope.

Solar Cooling in Colorado

At Fort Collins, Colorado, there are three solar houses built in a program directed by Colorado State University, described in Chapter 4. Two of the three houses are cooled during the summer by 3-ton lithium bromide absorption systems made by Arkla-Servel, but the solar collectors on these two houses are entirely different. One uses water-type panels with Olin Brass Roll-Bond alumi-

114

num sheet with integral channels coated with non-selective black paint and double glazing of glass. The actual mixture of antifreeze and water carries heat through a heat exchanger to a 1,100-gallon steel tank. Both summer cooling and winter heating are supplied by a fan-coil system. Power for the cooler comes from sun-heated water from the storage tank or directly from the collector at temperatures from 170° to 200° F.

On the other installation, called Colorado State University Solar House III, are panels consisting of glass tubes made by Corning Glass. These vacuum tubes contain copper strips to which are attached copper tubes, both strips and tubing coated with a selective black. The heat transfer fluid is antifreeze and water. With these evacuated tubes, described fully in Chapter 3, there is almost no heat loss from the solar collector strips and tubes, and liquid temperatures as high as 350° F are obtained. Liquid from these unusual glass tube panels heats a storage tank with 1,100 gallons of water. The remainder of the cooling and heating cycle is similar to methods used in the other Fort Collins house. A comparison of the efficiency of the two types of collector panels during the next two years will be interesting to architects and builders. It appears that both kinds of panel can be mass produced to achieve costs as low as $3.00 per square foot, so that either one or both will be highly useful as roofing material in the near future.

Georgia School with Solar Air Conditioning

In Atlanta, Georgia, there is a total of 10,000 square feet of solar panels at a 45° slope placed in twelve rows on the roof of the George A. Towne Elementary School. They are part of a system designed by Westinghouse Special Systems with performance monitoring by a team from Georgia Institute of Technology. The 576 solar collector panels are of the PPG type, described in Chapter 3, but there is an unusual feature. Next to each row of collector panels is a reflector made of one sheet of aluminized Mylar, a plastic, held by a bonding material, between two Mylar sheets; this sandwich is then bonded to ⅛-inch masonite. These reflectors perform two important functions: they increase the solar radiation striking the collector panels by 70% in winter; and in summer, they help to shade the roof from the worst heat.

This panel system will provide about 60% of the heating requirements for the school as well as all of its hot water. Storage is provided by three 15,000-gallon, insulated, underground, steel tanks storing hot water at temperatures up to about 135° F in midwinter. In summer, the temperature of the water from the collector panels is 200° F. It's hot enough and has sufficient energy from the large solar array to operate a 100-ton Arkla lithium bromide absorption chiller. During the summer one of the three 15,000-gallon storage tanks contains water cooled by solar heat so that there is always a chilled reservoir to maintain the school's cooling system, whether early in the morning or late in the afternoon. When necessary, there is auxiliary heating of the lithium bromide generator, but solar heating takes care of the majority of the air-conditioning load.

Maryland School's Solar System

At the Timonium Elementary School near Baltimore, Maryland, the middle wing has 5,000 square feet of solar panels. There is double glazing and an aluminum honeycomb support between the two panes, with another aluminum honeycomb below the lower glass and above an aluminum sheet painted flat black. The lower honeycomb is slotted so that water can run down the black absorber sheet. The assembly is bonded with epoxy, and has a back insulation of 1½-inch polyurethane foam, with a rubber strip at the edges of each 7 foot by 4 foot panel. Weight of this panel, designed by AAI Corporation, is about 3 pounds per square foot including deionized water furnished by a Culligan system. All piping is aluminum with rubber tube connectors.

Solar-heated water is stored in a 15,000-gallon vertical cylinder of welded aluminum, with tank and piping well insulated with polyurethane foam. Heating or cooling is distributed to school rooms by a fan-coil system. Water heated by the sun in these panels reaches 180° F or higher in the summer and provides energy for a 50-ton York absorption cooler. If solar energy is not adequate for heating or cooling because of prolonged cloudiness, there is an auxiliary oil-fired steam boiler.

Home in Ohio Cooled and Heated

A demonstration house in Columbus designed by a team from Ohio State University and built by Homewood Corporation uses PPG panels and obtains solar cooling and heating as well as hot water. A total of 800 square feet of collector area on three 45° sloped sections facing south is adequate to keep 2,200 square feet of living area at comfortable temperatures throughout the year. There are two 2,000-gallon water tanks for stratified storage—bringing hot water in at the top and taking cooler water from the

116

bottom—and a heat pump to aid in both cooling and heating. The roof panels contain anti-freeze which delivers heat to the water tanks through a heat exchanger. This solar-heated water powers a lithium bromide and water absorption system. The entire installation contains a small computer to keep track of system efficiency.

Based on the successful operation of this solar home in Ohio, the builder, Homewood, is planning to offer a series of single-family designs at prices ranging from $30,000 to $50,000. These houses will contain some back-up heating but it's expected that at least 70% of the requirements for space cooling and heating, and all domestic hot water, will be furnished by solar energy.

Ranch House in Arizona

In the Decade 80 Solar House in Tucson, Arizona, a 3,200-square-foot ranch-type home has been sponsored and financed by the Copper Development Association. Panels are of the type designed by Revere Copper and Brass described in detail in Chapter 3. Antifreeze and water are circulated at about 20 gallons per minute through the collectors, which comprise 2,200 square feet. Liquid from 1,800 square feet on the roof areas facing south and slanting at 27° from horizontal provides heat, through an exchanger, to a 3,000-gallon cylindrical tank. For domestic hot water, the supply either comes directly from this tank or, if necessary, passes through an auxiliary heater after solar preheating.

Cooling in this Tucson house is provided by two Arkla-Servel lithium bromide absorption coolers powered by water, which is heated in the summer by the copper solar panels to temperatures up to 220° F. If the water in the storage tank is not hot enough to operate the absorption coolers, an auxiliary heater cuts in—the same one used for standby heating of domestic water. A similar system applies for space heating in the winter. There are 400 square feet of solar panels on a separate building, with no glazing over the copper sheets because high temperatures aren't necessary. Water for heating the swimming pool is circulated through these panels to warm the pool when it's desirable. Other times, when the pool is warm enough, these panels are cut out of the system by suitable valves.

Apartment Houses and Commercial Buildings

A leading manufacturer of the Arkla-Servel type absorption air conditioners has made a number of 3-ton and 5-ton units which perform satisfactorily when the heat supplied is in the range of 180° to 220° F.

117

Since there are numerous solar collector panels which will deliver water or an antifreeze mixture heated to such temperatures, these Arkla units are now a viable element in home air-conditioning systems. This is particularly true because, during the hot summer months, a properly designed solar collector system will reach temperatures of around 200° F before 10:00 a.m. and its heated liquid will operate the Arkla cooler throughout the day. For standby service during prolonged periods of cloudy or rainy weather, an auxiliary gas burner is desirable. But solar energy will probably take care of 90% of the cooling load.

For apartment houses and commercial buildings, Arkla furnishes 25-ton absorption coolers designed for use with hydronic systems. That is, the cooler chills a water supply in pipes; then, for each pair of apartments or stores, a fan-coil unit blows warm air from the rooms over a coil of cold pipes, cooling the air, and returning it to the rooms.

HEAT EXCHANGE COOLING FOR AIR OR LIQUID SYSTEMS

Rock Bins

In the solar houses designed and built by Harry E. Thomason in Maryland, rock bins filled with stones are used to store heat in the winter and cold in the summer. During the winter, solar heat warms trickling water flowing down his collector panels; this water heats a storage tank, which, in turn, heats the stones packed around the tank. Then air is circulated by a blower through the hot rocks to heat rooms in the house.

During the summer, the sixth and most recent Thomason house uses a standard 28,000 Btu air conditioner during the coolest part of the night. For about six hours, from 10 p.m. to 4 a.m. (while the utility load is lightest), outside air is brought in through this inexpensive air conditioner to chill the rock bin in the basement. Then in the heat of the day, warm humid air from the house is blown through the cold dry stones. Rooms are cooled and dehumidified. According to Thomason, this approach doubles the efficiency of the relatively small air conditioner.

One of the three solar houses in the Colorado State University program at Fort Collins uses air-type collector panels (see Figure 46). Radiation passes through two panes of glass spaced ¾ inch apart and is absorbed by a flat black coating on galvanized steel. Beneath this collector is an air space connected to ductwork and a blower system which brings the air to a bin with 15 tons of stones

Figure 46. Colorado State University's Solar House II is the first solar home to use Solaron air-type collectors for cooling as well as heating.

about the size of ping-pong balls. When rooms need heat, a thermostatically controlled blower sends air through the warm bin to other parts of the house.

For summer cooling in this home, water is sprayed into a stream of air pumped into the building at night at about ground level. Evaporation of the water cools this air down to nearly the dewpoint, or about 50° F. This chilled air is circulated through the stone bin and cools it all night. Then during the heat of the day, on thermostatic command, warm room air is circulated through the bin, cooled, and returned to keep the house at a comfortable temperature.

In Figure 47 you can see how this cooling system works. It has been successfully developed and tested by Dr. George O. Löf, vice president of Solaron Corporation, Denver.

For cooling there is a fan, perhaps with water spray, to bring in outside air (6) from ground level at night to chill the stones in the bin (2). If it is necessary, an air conditioner can be placed at this point for further chilling of the night air, as in the Thomason house.

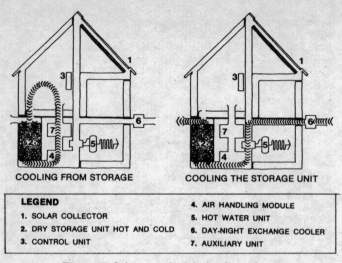

| COOLING FROM STORAGE | COOLING THE STORAGE UNIT |

LEGEND

1. SOLAR COLLECTOR
2. DRY STORAGE UNIT HOT AND COLD
3. CONTROL UNIT
4. AIR HANDLING MODULE
5. HOT WATER UNIT
6. DAY-NIGHT EXCHANGE COOLER
7. AUXILIARY UNIT

Figure 47. Solaron method for cooling a house.

Module (7) is an auxiliary heating unit which is cut into the system after prolonged cold, cloudy or rainy spells.

Other Examples of Rock Bin Cooling

Solaron Systems in Colorado

Among numerous houses using air-type solar collectors is one of the oldest solar homes in the United States. It is occupied by the original designer of this type of solar system, Dr. Löf, whose 2,000-square-foot home was built in 1959 (see Figure 16). On the roof are 600 square feet of collectors; there are 12 tons of granitic rocks in two bins. Other houses using the Solaron air collector panels include a house in Denver owned by an architect, Richard L. Crowther, which was retrofitted with roof panels, a rock storage bin and air circulation system; and a house in Wellington, Colorado, owned by the Perna family. In both houses, summer cooling is by means of air circulated through stones which have been chilled by night air. The air is pumped in at ground level and brought through a water spray when there is a succession of hot dry days.

Although architect Crowther has recently built another house in Denver using liquid-type collector panels and a 1,000-gallon water tank for heat storage with fan-and-coil space heating, the cooling is provided by air intakes at ground level on the north side of the house. There is a vertical duct from these intakes to the roof ridge so that cool night air travels through the house by convection. That is, since warm air rises, the cool air is sucked in to take its place. This intake of cool air can also be achieved by a fan.

There are some advantages to air-type collectors of the type made by Solaron in that the solar panels themselves may be less expensive, and there are no worries about leaks in the system. Ductwork made of fiberglass is reducing the cost of air-conditioning installations. So for space heating and cooling in areas where cool outdoor air can readily be collected at night, it's possible that a house using solar panels of the air type will save you money.

Sunworks House in Connecticut

Among those who are making both types of collectors is Everett M. Barber. He is president of Sunworks, in Guilford, Connecticut, which is making two types of solar panels utilizing an antifreeze mixture as described in Chapter 3. However, Sunworks also makes an air-type panel which Barber is supplying for a commercial purpose: furnishing sun-heated air for drying clothes in a laundromat.

On his own home in Guilford, Barber uses Sunworks panels containing water with 40% antifreeze circulated by a $1/12$-horsepower pump. For heating, Barber employs a 2,500-gallon storage tank and a fan-coil system. Heat is also stored in a 2-foot-thick layer of stones under the floor of the house. On sunny days, the stones are heated by hot air from under the roof, where the solar collectors are providing liquid temperatures up to 150° F. For cooling, in summer, the layer of stones is cooled at night by forced-air circulation, thus helping to keep the house comfortable during the day.

Cooling with Heat Pumps

Sol-R-Tech House in Vermont

A good example of how heat pumps can be used effectively in winter and summer as part of a solar air-conditioning system is the installation designed by Sol-R-Tech at White River Junction, Vermont. This is the Custom Leather Boutique building, described in Chapter 4 (Figure 38). With a floor area of 2,100 square feet, the boutique uses 910 square feet of Garden Way liquid-type collector panels (described in Chapter 3).

121

During the winter—in fact, about eight months of the year in this New England climate—solar heat from the panels warms water in a 5,000-gallon tank. The panel water contains a corrosion inhibitor but no antifreeze, and the liquid is drained into the tank at night. Three York heat pumps take energy from the tank water and provide hot air for heating the building.

In the summer, the heat-pump cycle is reversed. Hot air is pumped out of the rooms during the day and cooled by contact with the water in the storage tank. This water has been cooled at night by being circulated through the solar panels.

California Two-Story House

In Valley Center, California, there is a large two-story house with solar engineering provided by Schultz Field Enterprises. The collector is part of the roof and consists of one large panel, 36 feet wide by 14 feet high on an 18.5° sloping roof. Black-painted aluminum absorber plate backed by fiberglass insulation transfers solar heat to ⅝-inch blackened stainless steel tubes arranged along the 14-foot slope. These tubes are 5 inches apart and contain just water in this warm, Southern California location. The cover is a single layer of reinforced fiberglass coated with Tedlar. A differential thermostat turns on a circulation pump whenever the temperature of the water in the solar panel is higher than the temperature of water stored in two 1,200-gallon tanks underground.

For space heating, a blower blows air through a coil fed with hot water from a storage tank. Domestic hot water comes from a water-to-water heat exchanger in the tanks; there is also an auxiliary 40-gallon electric hot water heater. Auxiliary heating for the rooms is furnished by a 3-ton heat pump. In the summer this heat pump cools one of the large storage tanks through a refrigeration system, and this chilled water is fed to the fan-coil system. Solar energy provides about 80% of the heating and hot water requirements of this 3,600-square-foot home.

Advanced Photovoltaic System

Certainly one of the most advanced solar homes is the Solar One house that has been built by the Institute of Energy Conversion of the University of Delaware, Newark, Delaware (see Figure 48). It has several innovative features.

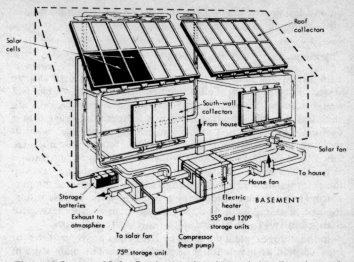

Figure 48. Layout of Solar One house using photovoltaic cells and eutectic storage bins.

Collector Panels Using Solar Cells

There are collector panels on 720 square feet of south-facing roof. These panels contain solar cells of the photovoltaic type (discussed in Chapter 11) which convert solar heat into electrical energy. The specific cells used on this Delaware house are of cadmium sulfide. In considering the solar heating and cooling of the building, the important fact of construction is that the cells form the back surface of a sandwich. Above the array of solar cells on each 8-foot by 4-foot panel is a ⅛-inch space filled by dry gas with a double cover, first of glass, then a top glazing of coated Plexiglas.

Solar heat warms the cadmium sulfide solar cells, which generate electrical energy. To cool these cells, and warm the building, air is blown through an insulated space 2½ inches deep behind the solar panels. This air keeps the solar cells from overheating while it is blown toward the peak of the roof at a rate of 3,600 cubic feet per minute, then piped to the basement thermal storage system.

There are other air-type collector panels on the south wall of this house. These are more conventional and do not contain solar cells. There is an outer sheet of coated Plexiglas, then a 1-inch air space and an aluminum sheet painted with a selective black coating. A 1,200-cubic-feet per minute (cfm) blower drives air upward

123

behind 130 square feet of these solar panels, and again the heated air is returned to the basement storage area.

Sophisticated Eutectic Storage

In this basement is an unusually sophisticated eutectic storage system designed by Dr. Telkes, a noted solar scientist who helped design some of the M.I.T. solar houses. These eutectic salts are chemicals having a melting point in a temperature range that is useful for home heating and cooling. Such a chemical, when it melts and goes from solid into liquid state, absorbs a considerable amount of heat. Meanwhile it does not change temperature. When the chemical compound returns to its solid state, it gives up the absorbed or stored heat; and again it stays at the same temperature. Thus, these eutectic compounds can be quite useful for storing and releasing heat. And they occupy considerably less space than a bin full of rocks that does far less thermal work.

In the storage system designed by Dr. Telkes are three stores, or bins, containing plastic tubes of eutectic salts. The main store in the basement has 8,000 pounds of sodium thiosulfate pentahydrate, which has an operating temperature of 120° F. Energy is fed to this bin by hot air from the roof and wall collectors to melt the eutectic salt, and store solar heat during cold weather. A separate stream of air, driven by a fan, collects heat from this store as part of the eutectic material changes back to its solid state.

There is a supplementary store, also in the basement, with 1,400 pounds of sodium sulfate decahydrate. This eutectic salt has an operating temperature of 75° F and thus it provides useful storage when the solar radiation doesn't warm the air behind the collector panels to 120° F. Heat is transferred from this smaller store to the main bin by means of a 3-ton York heat pump, which raises the temperature during the transfer by 45° with good efficiency.

A third bin of eutectic salts, called the "coolness store," contains 2,600 pounds of sodium chloride, sodium sulfate, ammonium chloride, borax and some additives. Its operating temperature is 55° F. During the summer, there is enough radiation from the panels to the night sky so that air passing behind the roof panels is cooled and can be used to chill the salts in this store. (Because atmospheric moisture impedes radiation, this system works best in a dry climate.)

Then a separate stream of air, circulated through the house, gives up its heat as it passes the third bin and the cooled, solid salts are melted. On cloudy nights, and nights when the outside temperature is greater than 75°, the York heat pump is used to take

124

energy from the coolness bin and discharge it into the 75° bin. That is, the eutectic salts in the 55° bin give up energy and solidify because of the heat pump's action at night. Then, in the heat of the day, this chilled bin cools the room air flowing over it as the salts are gradually liquefied.

Future Solar Air Conditioning

While the system of storing solar energy for both heating and cooling using eutectic salts is under development, there is a completely different technique being studied by the Institute of Gas Technology (IGT) in Chicago. It is called Solar-MEC (Munters Environmental Control) according to an article by Erik H. Arctander in *Popular Science*.

Patented by Carl Munters Company of Stockholm, Sweden, in 1950, this novel system is beginning to look useful because of research performed by Gas Developments Corporation, an IGT subsidiary also located in Chicago. Availability of a commercial Solar-MEC unit appears to be several years off, although Gas Developments is now trying to find manufacturers to build this efficient machine for heating and cooling buildings.

The basic principle of the MEC is to transfer heat and moisture between two chambers by means of two special wheels which turn slowly. One wheel has an aluminum honeycomb and acts as a heat exchanger. The other wheel contains an asbestos paper honeycomb impregnated with a dessicant, powdered aluminum silicate; it's called the "drying wheel."

In the cooling cycle, warm air from a house is pulled through the drying wheel turning at ⅓ rotation per minute (rpm). This wheel absorbs 80% of the moisture in the air. Because of this condensation, temperature of the dry air rises from 80° to 160° F. Now this hot air blows through a heat wheel rotating at 7 rpm and cools back down to 80°, then passes through water pads, giving up still more heat while evaporating some of the water. Output air to the house is wet and cooled to 55°. It mixes with room air and warms to approximately 75° as its humidity drops to around 50%.

The other portion of the cooling cycle serves only to regenerate the Solar-MEC unit so that it can continue to function efficiently. Air from outside the house is pulled in through the water pads in Solar-MEC and loses heat by evaporating some of the water. Now colder and wetter, this air gets some heat from the aluminum-honeycomb heat wheel and additional heat from a coil containing water heated by the sun. It's highly desirable, according to William F. Rush of IGT, that the water from the solar collector panel be as hot as possible, preferably at least 230°. Hot air, so heated, passes through the drying wheel, removes part of its water by evaporation, and then is blown outside.

With solar-heated water at 230°, Rush estimates that 80% of the gas required to heat the unit will be saved. As even more efficient collector panels are developed, such as the Corning Glass panels and others, the solar efficiency should be still greater.

As you saw in Chapter 3, if you keep the fluid circulating under the constraint of a watertight panel and plumbing on a hot day, it should be possible to reach fluid temperatures over 250° with several kinds of panels. Since those hot, clear days are the ones on which room cooling is most urgently needed, this higher temperature from solar heat is a real boon.

In the heating mode, the Solar-MEC is operated somewhat differently. Room air passes through the drying wheel, now speeded up to 10 rpm to make it a better heat exchanger. This house air gains heat supplied to the drying wheel from the solar-heated coil, passes through the stopped heat wheel, and then out into the home's ductwork system. For regeneration, outside cold dry air goes through to be heated by the solar coil, and carries this heat to the drying wheel, evaporating some of its moisture. The outlet delivers cool dry air to the outside.

There are other developments under way by various research organizations which promise improved ways to use solar heat for efficient cooling of buildings. Examples are novel Stirling-cycle and Rankine engines with energy provided from a solar collector. In general, such engines require much higher temperatures than those obtainable with flat-plate collector panels. Parabolic reflectors focusing the sun's radiation on a suitable heat exchanger (discussed in Chapter 10) may be the answer.

Meanwhile it's quite possible to design a home, apartment house, school or commercial building and get a substantial portion of the energy needed for cooling, as well as for heating, from a correctly installed solar system. With the kinds of equipment described in this chapter, it should be feasible to pay for the additional cost of installing collector panels, thermal storage and distribution equipment within ten years or less. In fact, by careful choice of the most efficient and economical designs, you may well be able to pay for your solar system in reduced fuel bills in five years. And then your house has become a truly economical place to live.

7 / Hot Water for Cabin or Trailer

If you happen to have a vacation home in some remote spot—in the woods, mountains, desert or beach—where there is a source of running water but no power or gas lines, you can still enjoy abundant hot water by installing a simple, inexpensive solar system.

The basic principle is simple and has already been discussed briefly in Chapter 5. Hot water, like hot air, rises. Although the water loses some of its energy while overcoming the force of gravity, if distances are kept relatively short this is no problem. The phenomenon of heat causing water to rise in any confined space, like a pipe, is called thermosiphoning and is very useful indeed when building a hot water system. Here is how to do it.

INSTALLATION OF A SOLAR WATER HEATER

Figure 49 illustrates a practical way in which to build your own hot water supply using solar energy as the only fuel. Cold water from the supply source is fed into the bottom of a 5-gallon container in which there is an ordinary toilet float valve to control the level. From this container the cold water goes into the bottom header of a solar collector panel.

In the panel, the water is heated and rises to the top header. From here, the water travels up by thermosiphoning to the 30-gallon storage tank. As indicated in the drawing, the sun-warmed water is brought in at the top of your tank through a tee and from the tank it is delivered to your hot water pipes for shower, kitchen sink, and basin. There is a vent pipe for exhausting hot air or steam from the system. This must be above the 5-gallon water reserve level, or else water will run out of it. A return line from the storage tank joins with the cold water pipe at a tee as indicated.

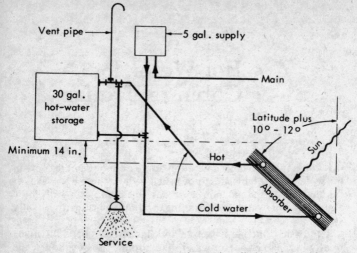

Figure 49. Layout of solar water heater installation for a cabin.

Components of a Solar Water Heater

A solar water heater is useful where you have a source of running water—whether from a main or cistern or spring—but no convenient electric or gas lines. It can be made from readily available materials, using hand tools.

Collector Panels

Your solar collector panel may be one made entirely by yourself. Or you may want to assemble your panel from components made by several manufacturers, or put it together at the site with materials and instructions furnished by a single manufacturer. See Chapter 3 for descriptions of available methods. It might also help to refer to Appendix 1, where there is a listing of companies furnishing solar systems, components and/or technical information useful for anyone building a homemade solar heater.

If you want to heat water in a larger tank than the conventional 30-gallon variety, it would be desirable to build a collector panel using materials which will give you considerable efficiency. For example, the SolarSan collector developed by William B. Edmondson is a good design. Or you can buy completely or partially assembled panels, as well as instructions for putting them together and installing your water heater, from many of the companies in Appendix 1, such as Corning Glass, Garden Way Laboratories, PPG Industries, Piper Hydro, Revere Copper and Brass, Reynolds Metal, Solar Power, Sunwater Company, Sunworks, and Tranter.

For a smaller ready-made system, the panels made by Hitachi and Sunsource (described in Chapter 5) have proved satisfactory in thousands of installations in Japan, Israel and other countries. Another convenient kit is put out by Sol-Therm. This kit includes two ready-made collector panels, one 32-gallon storage tank, all the pipes and nipples needed to connect the tank to the solar panels, plus a mounting frame so that you can install this heater on a flat roof or on the ground. Or, you can get a complete set of instructions from Solar Water Heater Company, which no longer builds solar panels but has installed hundreds of solar water heaters in Florida.

One of the most efficient solar panels that you can buy, already assembled for installation, is the solar heat collector made by Garden Way Laboratories, Charlotte, Vermont. Although these panels have been installed primarily in homes and commercial buildings where solar heat is an auxiliary, and there are pumps and differential thermostats as part of the system, there's no reason why you shouldn't use one or more of these panels at your vacation home. In fact, if you don't want to build a design such as SolarSan or one of the others in Chapter 3 by hand, the Garden Way store-bought unit is one good alternative. Its weight is only 25 pounds (without liquid) or just over 1 pound per square foot. This means that the collector can be mounted almost anywhere without putting stress on the structure.

Piping

The pipe you use in this system may be polyethylene or PVC plastic, copper, or galvanized steel. In general, you'll find it easy to work with plastic piping, but this is a matter of choice. Your hardware or plumbing supply dealer will be able to furnish you with both materials and advice as to how to use them, including how to install such fittings as tees and valves. Also see the information in Chapter 4.

There are a few precautions worth noting as you connect the plumbing for your solar thermosiphoning water heater—or any other solar system using liquid for heat transfer.

Keep your runs of pipe as short and straight as possible so as to make the flow of water easy, with minimum resistance created by your piping design.

129

When connecting a piece of pipe or any fitting, be sure that the inside surfaces are clean. Any dirt or chips will cause an obstruction that could cause trouble later.

Be generous in the diameter of pipe you use. This will make the siphoning action better. If your hot water storage tank is more than 5 feet from the top of your collector panel, treat yourself to 1-inch pipe. If the distance is more than 30 feet use 1½-inch diameter pipe.

Install a drainage valve in the cold water return line. You will want to drain the system if your vacation cottage is in an area where there might be a prolonged spell of freezing weather. And you will probably want to drain the system at any time when your family will be away from the cabin for weeks or months.

Since the bottom of the hot water tank must be at least 1½ feet higher than the top of the solar panel, you'll be running your hot water supply line up from the panel at an angle. Because your hot water must run uphill, here's one place where you may have to use larger pipe than you expect. If 1-inch diameter doesn't work, try 1½-inch or 2-inch pipe. You should get good siphoning action as soon as the sun warms the water in your panel. Also, if you use PVC pipe, it must be insulated if there's any chance of an unexpected freeze. Actually, as you may suppose, it makes good sense to insulate your hot water pipes, whatever material you've chosen.

Be sure that the air vent pipe extends above the top of the cold water reservoir, the small 5-gallon tank.

Hot Water Tanks

There are many kinds of small drums, pails and other metal containers which are suitable for the 5-gallon cold water tank. Figure 50 illustrates a tank made from a paint can. Whatever you use, it is important that this unit have a lid to keep dirt out of your system. Also, you will want to install a float valve of the kind used in toilets. A plastic valve from your nearest hardware store will do the job of automatically maintaining the water supply in this small reservoir, permitting flow from your cold water main when it's needed to replenish the supply in your solar panel.

Because the float valve must work every time, be sure to install the 5-gallon tank vertically. Otherwise you might have your float sticking against the side of the tank.

This float valve is tightly threaded into the cold water inlet pipe, brought into the reservoir through a water-tight, gasketed fitting. A similar fitting is used for taking the cold outlet pipe out of the tank.

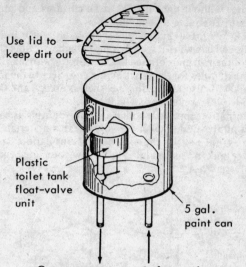

Figure 50. Five-gallon cold water tank with float valve.

There are many manufacturers of water storage tanks which can be used for your 30-gallon (or larger) hot water reservoir. Among the leaders in this field are:

Ford Products Corporation, Valley Cottage, New York

W. W. Grainger, Chicago

Norris Industries, Los Angeles

A. O. Smith Consumer Products, Kankakee, Illinois

Westerman Construction Company, Bremen, Ohio

Wood Industrial Products Company, Conshohocken, Pennsylvania

Most of these firms have wholesale distributors and dealers throughout the United States. Again your hardware or plumbing supply dealer will be glad to help you. And, of course, the large department houses and mail order firms stock hot water storage tanks of this kind.

Tank Insulation

It is extremely important to insulate your hot water storage tank. The better your insulation of this tank and the pipes carrying hot water, the longer they will retain heat. You can't have an effi-

131

cient heater without tight fittings at all closures and pipe joints, and without adequate insulation.

Typically, with a 30-gallon tank set horizontally as indicated in Figure 51, you can wrap it with 2 or more inches of fiberglass insulation. For the two ends of the tank, cut circles of insulation material, with exit holes for the pipes at one end. Secure the insulation to the tank with a suitable adhesive, such as epoxy, and then wrap the outside with weatherproof glass tape.

An alternative approach which is more durable is to build a large box of plywood, redwood or any durable material. Line this box with as much as 4 inches of fiberglass insulation. Then place your water tank in this insulated box. Now your sun-heated water will really stay hot.

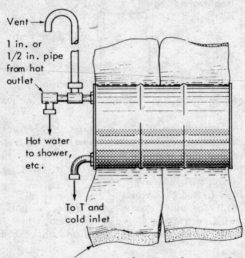

Wrap hot-water storage tank with insulation; ends too

Figure 51. Plumbing connections and insulation for hot water storage tank.

Installation of Solar Panels

If you've read earlier chapters, you know that your solar collector panel or panels should face south in order to get the maximum amount of useful sunlight throughout the year. Also, as discussed earlier, the angle of inclination is important. If you're using your vacation home only during the summer, then the tilt of your solar panel may be as much as 10° less than your latitude. For example, in the Los Angeles area with an average latitude of about 34°, the collector panel for a summer-only cabin should be inclined at from 25° to 30° from the horizontal. For year-round use, the tilt angle is more efficiently set at about 45°.

It may not be possible to put your solar panel on a south-facing roof at the proper angle and still place the hot water tank at least 14 to 18 inches higher. Ideally, your hot water tank could be in some attic space and your solar panel on the roof nearby. If the only available roof area faces east or west, you may find it necessary to use two collector panels instead of one. You'll probably need about 50% more surface for any area facing more than 20° away from due south. That means that South by Southwest and South by Southeast are acceptable directions to get high solar efficiency.

Alternative Mountings for Collector Panels

Even if you have a flat roof, you can put your solar panel on an inclined frame as discussed in Chapter 2, so that you get at least 25° of tilt. Construction of such a rack is described in Chapter 8. If there is no suitable roof area, you may be able to install your solar panel as if it were an awning shading a south-facing window. If so —and of course this also applies to roof or yard areas—be sure there are no trees which will prevent sunlight from getting to your collector panel. Should you decide to make your solar panel act as an awning, it might be well to use a lightweight design. Included in Chapter 3 are the weights of various types of solar collectors in pounds per square foot. This information is also available from each solar manufacturer.

Still another alternative is to put your solar panel on a sturdy rack facing south at some point in your yard. Bear in mind, however, that you must locate your hot water tank at least 1½ feet higher than the top of the solar collector, as shown in Figure 52. Also, since you're not using a pump in this completely solar-powered cabin system, you must place this tank in such a way that your hot water will flow through your vacation house with some help from gravity.

Remember that the placement of the elements in this solar heater system is crucial for its proper functioning. If you build a tight system, position the panel(s) so that they get plenty of sunlight, and use adequate insulation around your solar panel as well as the hot water pipes and hot water storage tank, you will be

133

Figure 52. Four Sunwater collector panels mounted below a hot water tank furnish plenty of heated water in the mountains near San Diego.

pleased with the results. For no expense in fuel, you'll have plenty of hot water. With routine maintenance, your solar hot water system should last twenty years or longer.

Determining the Number of Panels You Need

For your benefit in determining how many panels you need in a solar hot water heating system, Garden Way has prepared the tables in Appendix 2. By looking through the list of eighty American cities in Appendix 2, you can find the city nearest to you and determine your panel requirement per person. To determine the conversion of pitch to degrees, apply a ruler to your roof. A rise of 10.1 inches over a linear span of 1 foot between eaves and ridge means a 40° pitch to your roof. At this point, you compute a correction factor for your roof to account for its latitude.

MAINTENANCE TIPS FOR SOLAR WATER HEATERS

If you're using aluminum collector panels and/or aluminum pipe in a system for your domestic hot water, be sure to test the pH—acidity or alkalinity—of your water. With a pH of 6 to 8, your aluminum hardware will last a long time.

If the pH is lower and the water is more acid, then you'll have to do one of two things. Either insert a simple water treatment unit between your cold water supply and the inlet of your solar panel, or switch to a system that does not use aluminum.

Many of the aluminum systems now in use have been in service for a long time and will continue to be useful because either the water's pH is within the limits mentioned, or the collector panels contain an inhibitor and/or antifreeze mixture so that the liquid in the solar panels travels through a closed loop, delivering its heat to a large storage tank containing potable water through a heat exchanger. In this latter case, the liquid in the panels is changed every two years or more frequently, just as you change the antifreeze mixture in your automobile cooling system.

USING YOUR FIREPLACE AS A SOLAR SUPPLEMENT

You can give your solar water heater a boost if you have a fireplace in your vacation cottage. This is particularly feasible if you're using copper, galvanized steel, or aluminum pipe for your hot water. If you can conveniently run a coil of piping through the firebox of your fireplace, you can take advantage of the heat of your fire to keep water hot in your storage tank.

One way to do this is to have a separate loop of pipe running from a tee at the bottom of the hot water storage tank, through the coil in your fireplace, and back through another tee to the top of your storage tank.

If you're using plastic pipe, you can still have a metal coil for the hot portion of the fireplace. Just use appropriate mating connectors between plastic pipe and a copper heating coil.

ALTERNATE DESIGNS FOR SOLAR WATER HEATERS

SAV Solar Cylinder

A novel design of solar collector, proved useful in heating water, is the SAV system, so named after its inventor, Stephen A. Vincze, a consulting engineer in New Zealand. Shown installed on a roof in Figure 53, these SAV cylinders are imported by Fred Rice Productions, in Van Nuys, California.

Containing 12 gallons of water, each SAV unit could act as its own storage tank. Construction of the cylinder is shown in Figure 54. Cold water enters through inlet (7) at the lower end of the unit

135

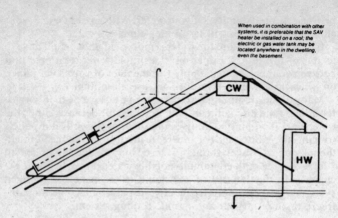

When used in combination with other systems, it is preferable that the SAV heater be installed on a roof; the electric or gas water tank may be located anywhere in the dwelling, even the basement.

CW

HW

Figure 53. Typical SAV solar water heater installation.

and circulates through the space between the black metal heat collector (1) and an inner cylindrical guide (2). There is dual cylindrical glazing with an outer glass shell (4) separated by an air space (6) from the inner glass shell (3), with another air space (6) between the inner glazing and the absorber. Sun-heated water comes out of the outlet pipe (8), above which is a vent pipe (9) needed to exhaust trapped air and steam from the system.

According to Rice, this SAV cylinder may be connected in series or parallel and is usually installed on a roof with a reflecting surface material under the solar collectors. An aluminum reflector with pressure-sensitive adhesive backing, available in 2-foot wide rolls, is supplied by Rice. It may readily be attached to any roof surface, has 90% reflectance of solar radiation, and is cleaned by washing with water.

Because of the curved surface of the SAV design, its reflector collects more insolation than flat plate solar panels. Thus, this unit is said to be more efficient in terms of the amount of water warmed to a given temperature per square foot of collector. In one test in Wellington, New Zealand, latitude 41° South, there was a comparison between Solapak copper flat plate units made in Australia versus the SAV cylindrical collectors. It was found that six Solapak units with a total area of 48 square feet (each panel 4 feet by 2 feet) produced 20 gallons of hot water up to 120° F. Under identical conditions two SAV cylinders with a heat collector area of only 6 square feet brought 20 gallons of water up to 160° F.

Even tilted only 13° from the horizontal, this cylindrical solar heater works rapidly if properly backed by a reflector on your roof

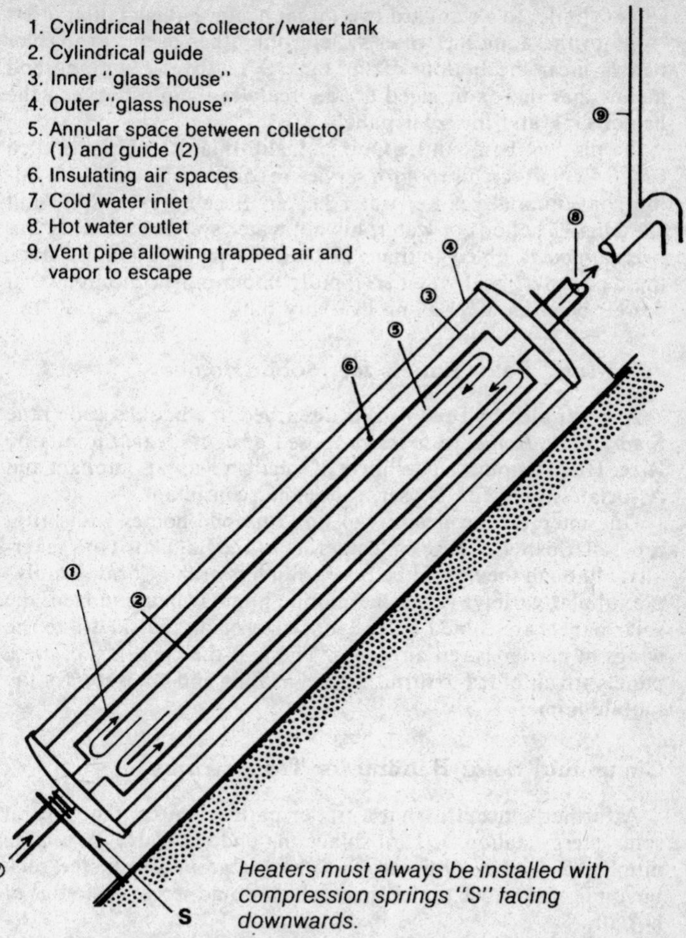

1. Cylindrical heat collector/water tank
2. Cylindrical guide
3. Inner "glass house"
4. Outer "glass house"
5. Annular space between collector (1) and guide (2)
6. Insulating air spaces
7. Cold water inlet
8. Hot water outlet
9. Vent pipe allowing trapped air and vapor to escape

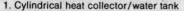

Heaters must always be installed with compression springs "S" facing downwards.

Figure 54. Construction of SAV cylindrical water heater.

or rack. Temperature rises averaging 46° F per hour have been measured with a SAV unit having this small inclination. When placed at a more favorable angle, the cylinder will do even better.

Its weight is 381 pounds, including water, and the price is $350 per 12-gallon cylinder. Typically the SAV unit is used as the major water heater, tied in with a gas or electric heater as an auxiliary. For use in a wilderness cabin, it's possible to connect this type of solar cylinder to a standard two-outlet heater having both outlets —one to the domestic water system, the other back to the solar panels—near the bottom of the tank. This saves plumbing and means that there's no need for a circulating pump between the heater tank and the solar panels.

On his own home in La Quinta, California, Rice has installed four SAV units on his roof for service in conjunction with a 50-gallon conventional gas hot water heater. Because each SAV unit contains 12 gallons of water, his hot water system's capacity has been almost doubled—4 times 12 plus 50 equals 98 gallons. Also, these cylinders heat water so rapidly under most conditions that there's been an 80% saving in utility bills.

"Flip-Top" Solar Panels for Mobile Homes

Another idea by Fred Rice is described in a booklet called the *Solar/Sonic Home.* In it are proposed designs created jointly by Rice; Hugh Kaptur, an architect of the firm Kaptur, Lapham and Associates; and Ted Poyser, a designer consultant.

One interesting concept is to build mobile homes with "flip-top" solar panels. These collectors are of the flat plate type generally, although they could be SAV cylinders or the Corning multiple tubular variety. When the mobile home is being moved, the solar panels are folded on the roof area or on the sides like the wings of carrier-based airplanes. Then, at the chosen site, these panels are unfolded to furnish space heating and hot water for the mobile home.

Communal Solar Heating for Trailer Parks

A further concept is that a trailer park might contain several solar energy stations to furnish heating and hot water for a large number of trailers. One idea is to have a ground-insulated rock bin underneath each trailer space which would serve for storage of hot air.

Although Rice's publication does not suggest any definitive hardware for this kind of a solarized mobile home estate, it's interesting to imagine how this idea could be made feasible at a reasonable cost. Here are some suggestions:

Space heating could be provided for numerous trailers by establishing several solar service stations of the A-frame type designed by International Solarthermics Corporation, described in Chapter 4. Then hot air stored in the rock bins contained in these A-frame buildings would be distributed through insulated plastic ducts to all the trailer sites. The hot air outlets would be capped when no trailers are at the sites. They would be connected into the trailer's ventilating system by means of a flexible plastic hose with a valve and a thermostat. This could be accomplished as easily as connecting the trailer to the park's water.

Hot water for trailers could be provided by one major service building in the trailer park with liquid-type solar panels on its roof. These could be any of the numerous types described in Chapter 3, or the SAV cylindrical units. These solar collectors would be connected to a large underground tank, heavily insulated and surrounded by a rock bin. In a cold climate, where winter temperatures are frequently below freezing, the liquid in the solar collectors would be an antifreeze mixture, and the heat from this mixture would be transferred through a heat exchanger to the water in the single big tank.

From this tank would be extended a loop of copper piping, such as that used in the Piper Hydro system (described in Chapter 4), so as to provide a hot water tap at each trailer site. To furnish auxiliary heating during prolonged spells of bad weather, there would be a gas-fired or electrically heated boiler in a small shed strategically located for each group of twenty trailers. In this way, every trailer would have hot water immediately available merely by connecting to the insulated hot water main.

8 / Heating Your Swimming Pool

Using solar energy to heat your pool is one of the easiest ways to make sunlight work for you. As you look unhappily at your monthly gas bill for pool heating, it won't take you long to decide that free sunpower may be a better answer.

This goes along with the objectives of municipal authorities. In New York State and many cities in California, if you want a permit now to put a pool in your backyard, you'd better use solar heating—or give a very good reason why not.

As most owners are aware, having a pool without some form of heating becomes intolerable for a family that likes to swim. It's a waste of backyard space and of your big investment in a pool if you don't keep it heated.

Even though the cost of heating your swimming pool by sunlight is initially somewhat higher than putting in a gas heater, the solar system pays for itself rapidly. If you install your solar heater yourself, as described in this chapter, the total price for materials will range from $600 to $1,000. You can figure your system will pay for itself in two to four years at present gas prices. And they are almost certain to go higher.

HOW SOLAR POOL HEATERS DIFFER FROM SOLAR DOMESTIC WATER HEATERS

Providing a heating system for your swimming pool is quite different from the solar heaters previously described in Chapters 4 and 5 for heating your home and its hot water, although it is possible to combine both home and pool heating, using some auxiliary gas heating.

In general, your home heating and hot water system utilizes solar panels designed to achieve water temperatures ranging from 140° F to 200° F. Heated water is stored in an insulated tank usually used as a reservoir for solar heat when the sun is not shining.

141

Rapid Circulation of Water through Collector Panels

When your object is to heat your pool, your only aim is to raise the temperature of several thousand gallons of water to a maximum of around 80° F. You circulate water through solar panels just as for home heating. But you move the water through the panels more rapidly and keep these energy collectors relatively cool. As a result, the panels work more efficiently.

As stated by Fafco, a leading maker of solar heating systems for swimming pools: "A solar pool heater can deliver a large volume of water only 2° to 5° warmer than it was when it entered the collector. Thus this kind of solar panel more than makes up in volume what it gives up in higher temperatures. Solar pool heaters can operate at efficiencies of 70-80%, whereas most solar hot water heaters operate in the 30-50% range."

Minimum flow in Fafco panels is calculated at 1½ gallons of water per minute per panel, and maximum is 8 gallons per minute through each panel. Absorption of solar radiation is rated at 95% over the spectrum from ultraviolet to long infrared. With an array of seven solar panels handling 50 gallons per minute, your pool pump will have to supply less than 5 pounds per square inch (psi) additional pressure. This increased pressure will appear on the gauge at your filter.

Use of Plastic Panels

Because collectors for heating swimming pools don't have to withstand high temperatures, they can be made of a durable modern plastic material. There is no need for a glass cover or the type of sandwich construction we have previously seen in home heating and cooling systems. In fact, assuming you're primarily interested in an efficient, economical solar system for your pool—particularly if you're going to make your own installation—the choice of plastic collector panels is logical. At the end of this chapter is a list of some manufacturers of solar heating systems for pools.

As you might imagine, the color of the plastic used in these heater honeycomb structures is black. Here's a case where black serves two basic purposes. It is ideal for absorbing a maximum amount of sunlight and converting it into heat. It also happens that the best materials, called additives, for keeping plastic pipes from deteriorating because of the ultraviolet rays in sunshine are carbon black and suitable antioxidants.

This is a lesson learned long ago by telephone and power companies. The material used for insulating their wiring is a plastic resin compounded with carbon black. It is a stabilizer which makes the compound opaque to both visible and ultraviolet light. Because ultraviolet radiation can't penetrate beyond a thin surface'layer of the black plastic, properly compounded material will last in outdoor service for twenty years or more.

Understanding these facts is more important in your selection of plastic solar panels for heating your pool. As an example, heater panels made by Fafco, which include ultraviolet inhibitors, have been in operation as long as the company—more than eight years—without appreciable deterioration (see Figure 55). And this includes installations in such differing climates as northern California and southern Florida.

Need for Auxiliary Heater

It is important to keep in mind, however, that while plastic collector panels are durable and less expensive than metal collectors, they are also less efficient. Therefore, don't expect your solar system to deliver tropically warm water throughout the year without

Figure 55. Nine Fafco plastic solar panels on rack double as an awning while heating large pool.

some help. For this you will need an auxiliary conventionally fueled heater. However, if you and your family will be satisfied with a cool pool during the winter months when the outdoor temperature is similarly cool, then the unassisted system will do your entire job.

For those who insist on having a pool warmed to 85° all year long, it's necessary to include a fuel-fired heater (usually gas-fired, although it might be electric) in the system. During cold months, this heater will boost the temperature of water going into the solar panels by about 10°. Even with some auxiliary use of this gas heater, you'll notice a saving in your utility bill for pool heating of about 70%.

An alternative is to use the more efficient and expensive metal solar collectors described in Chapter 3.

LOCATING YOUR SOLAR PANELS

In planning to assemble an array of black plastic panels to heat your pool, there are several factors to keep in mind.

Determining the Number of Collector Panels Needed

First is the size of your array. Evidently this will be a major consideration as to where you place the panels. Typical sizes of black plastic collectors include Fafco, with 8-foot by 4-foot and 10-foot by 4-foot units; Fun and Frolic, and Solarator, each 6 feet by 3 feet.

How many panels will you need for heating your pool comfortably? This depends on the size of your pool and the direction in which you face your assembly. Keeping in mind that you want to maintain a substantial flow of water through your heater panels, with the temperature relatively low to achieve greater solar heating efficiency, a rule-of-thumb proves to be: plan to use an area in panels equal to at least 50% of the surface area of your pool. In other words, with a 40-foot by 20-foot pool having 800 square feet of surface, you need at least 400 square feet of solar panels.

Determining Direction and Tilt of Panels

A second major efficiency factor is placement of your panel assembly. Ideally your array should face south in the Northern Hemisphere—and north if you're in Australia, Brazil, Chile or South Africa. (See Figure 56.)

Also, the degree of tilt, or inclination, of your array is important. As pointed out in earlier chapters, it is best to place your panels at an angle from the horizontal equal to your latitude plus 10°. Thus, in Southern California or Arizona, your solar collectors should be approximately 40° from the horizontal. In St. Louis the angle is 45°, and in New York about 50°.

Figure 56. Installation of six Fafco solar panels on roof facing south heats nearby pool.

Remember that so far we've merely considered ideal conditions. It's quite possible to install a satisfactory solar heating system for your pool even if your topography makes it impractical to have a perfect setup. You simply need to use more panels.

If you can place your array on a roof or a rack with the correct inclination for your latitude, here's a summary of solar panel requirements using plastic panels: With a panel assembly facing south, collector area should be a minimum of 50% of your pool area; with a panel assembly facing west, have a minimum of 75% of pool area; with panels facing east and west, try to put most of your panels on the west side to get the afternoon sun, and use a collector area equal to at least 75% of your pool area. With more efficient collectors like those in Chapter 3, you'll need half as many, but the cost is apt to be higher.

Panel Arrangements on Your Roof

As you can see in Figure 57, there are many ways in which you can arrange panel assemblies on your roof. You can place the entire array on one section of your roof, either as a single bank of panels or as two or more arrays. Or, it may be more practical to place your panels on two or more areas of your roof, or at several levels on your roof; on your house roof and garage roof, for instance.

Suppose your roof area is flat. You have at least two alternatives that will work. One is to set your solar panels in a rack having a tilt of at least 2° from the horizontal. This rack may be made either of metal or of seasoned and painted wood—a good choice is redwood because it's light, easy to work with, and durable. You need a tilt of a minimum of 2° to insure reasonable heating efficiency, although you'll need an increased panel area of at least 20% if your inclination of the array is 2° instead of 40°. Therefore it's worthwhile to build as much elevation for the upper end of your assembly as necessary to achieve the angles recommended for your area. Remember that in winter months the sun is relatively low on the south, and those slanting rays deliver far less heat than sunlight striking your panels vertically.

Another point to keep in mind for any rooftop installation is the weight of the solar panels. While black plastic collectors are far lighter than a metal-and-plastic sandwich construction, the weight of a typical panel filled with water is 1 pound per square foot. So if you need 400 square feet of solar panels, the assembly will weigh 400 pounds. However, this weight is evenly distributed over the area of the roof which it covers and should cause no problem on any soundly built home or garage.

Mounting Panels on a Rack

If there's no way in which you can place your solar panels on your roof, mount them on a rack in a convenient location.

How to Build a Rack

If you use light weight solar panels such as Fafco makes, a rack to support them is easy to build. Sketches in Figure 58 indicate a suggested method. After constructing an outer frame and legs of wood as indicated, you might cover this with corrugated fiberglass sheet to provide a continuous surface to support your panels. The height of the legs of the rack will depend on the amount of inclina-

146

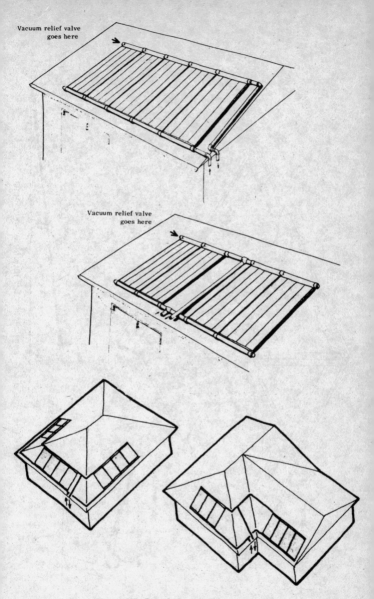

Figure 57. Panel arrangements on roof.

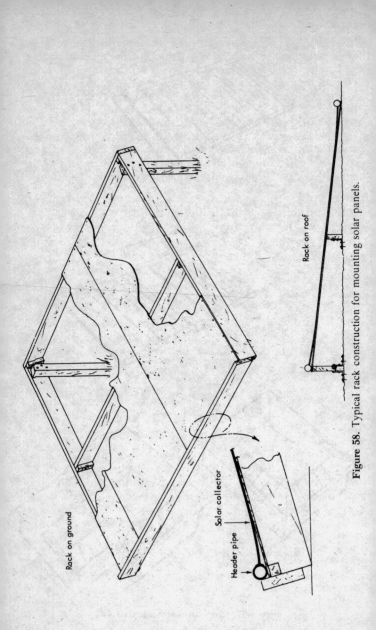

Rack on ground

Solar collector

Header pipe

Rack on roof

Figure 58. Typical rack construction for mounting solar panels.

tion needed in your latitude and the top of the legs will be cut at an angle reflecting this inclination. All this is on the supposition that your house or garage doesn't have a roof with a suitable pitch for placing your solar panels.

The cross-section view in Figure 58 shows a proposed approach if you want to put your solar heater on a flat gravel roof. The upper end of the array of panels rests on a fiberglass sheet supported by a 2-inch by 12-inch board, with another 2-inch by 6-inch board midway under the sheet. These boards are secured with steel L brackets as indicated; use care to seal these areas with a suitable caulking compound after installing the brackets so that your roof won't leak.

A rack-mounted solar heating system installed in January in Los Angeles immediately brought the pool temperature up to 70° and kept it slightly above that level until March. Then the pool reached an average of about 82°, the level at which the thermostat in the automatic control system was set. It will stay just above 80° until late October, depending on the weather, and gradually slide down to about 72° in midwinter.

INSTALLING YOUR SOLAR HEATING SYSTEM

Attaching the Collector Panels

For the type of plastic solar panel made by Fafco and others, it is most important that each panel be properly tied down to the roof or mounting rack. Figure 59 shows how this attachment is made. Nylon straps are used on the high or upper side of the banks of panels. You secure the bottom side of each panel with vinyl straps. This more elastic vinyl strapping allows for expansion and contraction of the panels with changes in temperature.

Of most importance in securing your panels is proper fastening of the nylon center-ties. If the panels are properly tied down, they will have a tendency to flatten down against the roof when the wind blows. If they aren't properly secured, your panels may begin to act like sails. Be sure that you fasten each panel firmly to clamps on the mounting surface (roof or rack) at every point indicated, including the center-ties between panels.

If you are installing the panels on a roof, use a durable sealant between the panels and roof. This sealant should be compatible with the roofing material on your home. For example, tar and gravel roofs must be finished with hot tar. With a wood or composition roof, any of the epoxy sealants available at a hardware store is suitable.

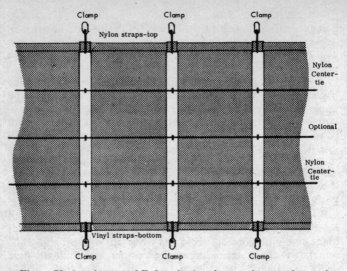

Figure 59. Attachment of Fafco plastic solar panels to roof or rack.

When you are initially installing an array of panels, tighten the clamps, or turnbuckles, one-third to one-half of the thread. After all panels are in place, most of the clamps should be tightened still further, to at least three-quarters of the total thread distance. The exception is at the end panels, where you are clamping against PVC plastic pipe or some other rigid material. Be sure to place clamps carefully on the couplers between panels, relative to the body of each panel, so that there is no skewing of couplers or panels.

Use an extra long set of couplers at the end points where the array of panels connects with the feed line from the pool (low end of the panel) and the return line to the pool (high end).

Connecting the Plumbing

Your array of solar heating panels is inserted into your pool's plumbing system in either of two ways:

Installations Without an Auxiliary Heater

If your solar heater is designed to do the whole job, as is the case with many installations in mild climates, then its plumbing connections are made with the input feed line from the filter and the output plumbing going back to the pool (see Figure 60). It is ex-

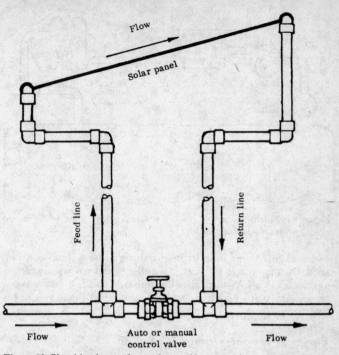

Figure 60. Plumbing layout for a solar pool heater without auxiliary heater.

tremely important to keep an unobstructed line carrying the heated water from the panels to the pool. Any blockage in this return line is apt to result in overpressure being applied to the panels. They are being filled at the input end. If the output is restricted, your panels may be damaged. And no manufacturer's warranty will cover this kind of negligent installation.

In pool systems with vacuum filters, the plumbing connections will be made to your solar array right after the pump, and then the return to the pool.

Installation Using an Auxiliary Heater

If you have a gas heater already, or plan to have one, then you connect your solar heater between the filter and the fuel-fired heater. This is diagrammed in Figure 61. By this technique you get the best results from both heaters. Water heated by the sun is returned through the standard heater, thus reducing the amount of

151

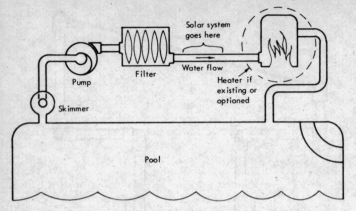

Figure 61. Fafco diagram of pool installation where solar heater is connected with an auxiliary heater.

fuel consumed even on cold days or during a long cloudy or rainy spell. During most of the year, it is unlikely that you will need your gas heater at all. With proper thermostatic control, your gas will be consumed at a very low rate because of all the help you're getting from the sun.

Ensuring Proper Water Circulation

Fortunately for the design of your system, you have a useful ally in your pool pump. It is made to handle high volumes of water at low pressure. This is one reason plastic solar panels are so practical. Also, your pool pump will easily force the water through the filter and then up to your roof, even if it is a two-story house.

Water flowing through the panels of your solar heater returns to your pool by gravity. This downward flow, in turn, creates a suction or siphoning effect which makes it still easier for the pump to keep the flow through the panels moving. In fact, this suction is so strong that solar heater systems include a vacuum relief valve so that the vacuum created by siphoning won't pull water out of the panels too rapidly.

When your vacuum relief valve is working properly, it's likely that bubbles will appear in the outlet of the return lines in your pool. Don't worry about these bubbles. They are your assurance of satisfactory operation of the system you've installed.

In laying out the plumbing connected to your solar heater, you connect the feed line coming from your pump and filter to the bot-

152

tom header of your panel array. The return line to your pool is connected to the top header of the panels. You install your vacuum relief valve in the plug of the top header in the panel farthest away from the feed and return lines.

Choice of piping

Recommended for use with your plastic heating system is standard PVC pipe, available at hardware, building and plumbing supply stores. This pipe is easy to cut and thread. Heat losses through this pipe for typical distances involved in a pool heating system are insignificant.

If the distance from your pool pump to the solar panels is less than 60 feet use 1½-inch diameter Schedule 40 pipe. If the distance is greater than 60 feet use 2-inch diameter Schedule 40 pipe.

Should your present pool system utilize copper pipe, then you'll need to solder on threaded copper fittings for adapting to the plastic PVC pipe used in the lines to and from your solar heater panels. This soldering is readily done with a small propane or butane torch. Solder on a copper male fitting because you'll screw on to this a female adapter made of plastic.

Feeding water to solar panels

There are two basic methods for feeding water to your solar heater whether it consists of one or several arrays of panels.

END FEEDING

End feeding is the most common method if you have a single assembly of solar panels. The water coming from the filter is fed into the bottom header at the end of the array closest to the pool. The water will fill the header of each panel and then begin to rise evenly through the panels until the top headers are full. Now the water flows out of the top header back to the pool.

One precaution is that you should always mount your panels in such a way that the header pipes, those at the top and bottom of each panel, are parallel with the horizontal. This is done to insure even flow of water through your entire array of panels.

CENTER OF TEE FEEDING

Center of tee feeding is often used if there are several arrays of solar panels. The water is fed into the panels by means of a tee fit-

153

ting which divides the flow of water and sends it to arrays remote from each other as illustrated in Figure 57. If you find that this feeding system produces uneven feeding between arrays, it is easy to remedy. Simply put a gate valve in the line feeding the group of panels receiving a surplus of water and reduce the flow to that array. A practical way to tell if you're achieving equal flow in separated arrays is to touch the surfaces of the panels. With this kind of a plastic-panel solar heating system, these surfaces should all be equaly temperate—no more than 85° F.

In multilevel panel installations, illustrated in Figure 62, you may wish to use either end feeding or center of tee feeding. In either case, separate arrays of panels should always be connected with your plumbing lines providing parallel rather than series flow. Efficiency of your solar heating system depends on a high flow rate and a relatively low surface temperature. When your panels' surface is cool, the heating efficiency is as high as 75%. Most of the solar energy reaching your system is being used to heat your pool.

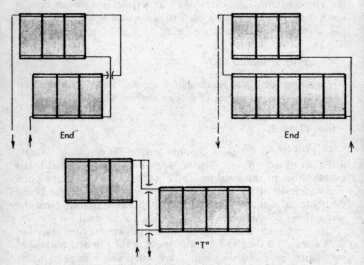

Figure 62. Plumbing connections for panel arrays on two levels.

In fact, if you should have a panel considerably warmer than the others, you may have a bad panel. This type of high-flow system, using plastic panels, returns water to your pool no more than 6° F warmer than your pool. If the return water is more than 6°

154

hotter, there may be something obstructing the flow and you should check your whole system. When the temperature goes up drastically it is because the flow has dropped. This means that you can feel your panels getting hotter, and there will be a smaller volume of water coming back into your pool through the return line. If this happens, try to find the obstruction in your system.

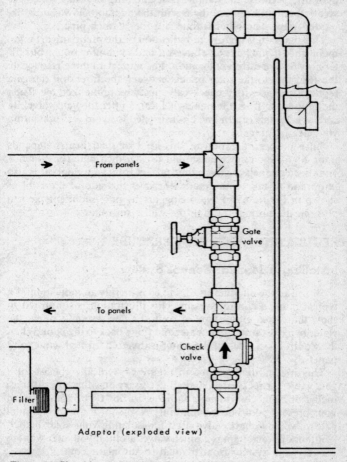

Figure 63. Plumbing connections from solar heating panels to an above-ground swimming pool.

Plumbing Arrangements for Raised Pools

A slightly different arrangement of plumbing is needed if your pool is above ground. See the diagram in Figure 63. The Fafco plumbing kit includes an adaptor to connect to the filter, a gate valve and check valve, all necessary tee and ell fittings, plus a plug to stop up the former through-the-wall return. From the existing pool filter, remove the flexible hose and long ell fitting which are screwed on, and replace them with the adaptor shown in an exploded view. The other fittings are for 1½-inch pipe.

This diagram is merely an indication of the plumbing arrangement since if you have an above-ground swimming pool, your filter may be in a different location. The important thing is to locate the feed line to the solar panels between the filter and the gate (control) valve, with a check valve just before the feed tee. Then the return line from the panels is located after the gate valve. Be sure to keep this return line unobstructed to avoid overpressuring your heating panels.

Since proper operation of this kind of solar heater depends upon high flow rates, you should use 1½-inch pipe (or 2-inch if your heater panels are more than 60 feet from your pool). It is also important to take your return line over the side of the pool as shown in Figure 63 to avoid clogging by pool sediments and to plug up the normal through-the-wall return line.

REGULATING POOL TEMPERATURE

Installing a Manual Control System

Shown in the diagram in Figure 64 is merely an indication of a typical manual control system. This illustration is simplified to show that when either type of control valve system is open, the water bypasses your solar heater and flows back to the pool. When this control is closed, all the water is diverted through your solar panels.

This system, diagrammed in Figure 64, typically consists of a gate valve, a check valve, and all necessary bushings and fittings, including connectors to mate with the size and type of pipe used in your present pool house plumbing. Also shown are an optional gate valve and check valve used for isolating your solar heater. Another optional item is a pinch valve, useful if you later want to convert your system from manual to automatic control.

Your first step, assuming you're installing your solar heating system for an existing pool, is to cut out enough pipe from your

156

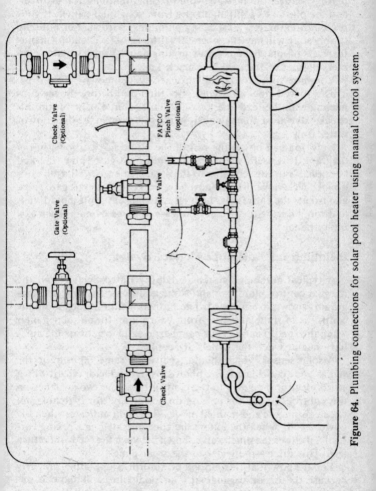

Figure 64. Plumbing connections for solar pool heater using manual control system.

present plumbing to make room for the feed and return lines as well as the plumbing between them.

Next step, as previously described, involves providing suitable mating connections between your present plumbing, if it's copper, and the plastic PVC piping to and from your solar panels. If your present plumbing is plastic, any plumbing, hardware or building supply store will have the necessary fittings. In fact, leading manufacturers of solar heaters for swimming pools furnish suitable couplers as part of their do-it-yourself kit.

Your next move is to install the check valve shown in Figure 64. This is placed right after your pool filter and before the tee connecting with the feed line to your solar heater. The arrow on this check valve must point away from the filter and in the direction of water flow.

Now you can install the rest of the plumbing as shown in the diagram. It is useful to insert the additional gate valve (marked "optional") in the feed line to the solar heater as well as the optional check valve in the return line. You can close this gate valve and isolate the entire solar heating system in case you should want to drain it during the winter or in case of some damage, such as a faulty panel.

Installing an Automatic Control System

A typical automatic system, made by Fafco, includes all the manual control plumbing and some additions. A simplified diagram is shown in Figure 65. The basic function of the automatic control is to route all water from your pump to the solar panels when the sun is heating your panel array, and to activate the auxiliary heater when the panels are cool.

A solar sensor, which should receive the same exposure to the sun as your panel array, is a photosensitive detector circuit which sends electronic signals to the control box. The two switches on this control box are for turning on power, and for switching between automatic and manual modes. The light indicates when the power is on. When the automatic control system is working correctly, it closes the pinch valve shown between the feed and return tees. This directs the water to the solar panels.

You'll know that this automatic control is operating properly because the pressure gauge at your pool filter will show an increased pressure of about 5 pounds per square inch. Also your panels will become cooler as the water flows through them, and you'll see bubbles in the inlet(s) to your pool as the air from the panels is first purged. To set this system up:

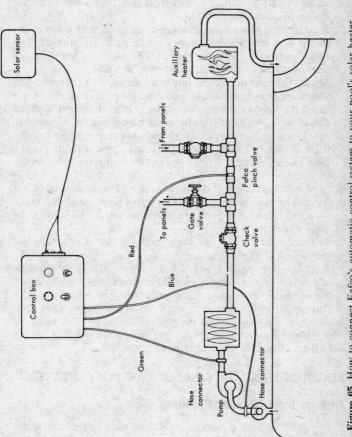

Figure 65. How to connect Fafco's automatic control system to your pool's solar heater.

1. The first step in installing the automatic control is to place the pinch valve as indicated. It can replace the manual gate valve, if you wish. However, most pool owners install both the manual gate and the pinch valve, leaving the manual valve open while the solar heating system is operating automatically. You should install the pinch valve before the control box and solar sensor are connected.
2. Next, drill $5/16$-inch holes in the feed lines on both sides of the pump—suction and discharge—as indicated in Figure 65. Tap these holes with a standard $1/8$-inch NPT pipe tap. Now you can screw into place the hose connector fittings.
3. Mount the control box securely to a wall or post near enough to your pool plumbing so that the tubes from this control unit can reach the hose connectors on either side of the pump and the pinch valve. With this control unit the tubes are color tagged so that green (power) goes to the pipe between the pump and the filter, on the discharge side of the pump; blue (exhaust) to the pipe on the vacuum side of the pump; and red (drive) to the pinch valve.
4. Now you can connect the solar sensor. This sensor does not have to be mounted near your array(s) of solar panels but make sure the sensor is exposed to the same degree of sunlight as your panels.
5. Install the check valve in a horizontal position prior to the inlet tee before the feed line to the panels, and a gate valve in the feed line. This permits you to isolate your solar heating panels if you're going to be away from home for a long time, or if you should have to repair a damaged panel that is leaking.

During any part of the year when you are not using your solar heater, you should exercise the pilot solenoid in the automatic control box. You accomplish this by flipping the switches on and off several times so that the light goes on and off.

MAINTENANCE OF YOUR SOLAR POOL HEATER

How to Spot Needed Repairs

Typically all major parts of a solar heating system are under warranty for a year and should last far longer with proper installation and routine maintenance. But little things can, or course, go wrong, and if not caught in time can lead to larger troubles.

Here, then, are some hints for troubleshooting this kind of a system. If the light on the control box fails to come on, make sure the power switch is on and the control is plugged into a working power receptacle. Then switch to "manual." If the light comes on, then the solar sensor needs adjustment or replacement. If the light fails to come on in the manual mode, check the fuse.

If the light is on, indicating proper automatic control, but water isn't being diverted to your solar panels, disconnect the line to the pinch valve. With the pump on and the light on, a stream of water should flow to the pinch valve. If it does, the pinch valve is defective. If it does not, the solenoid is defective.

As to leaks in your plumbing system, if water comes from a fitting, simply tighten it or replace it. If it comes from the control box, first unplug the power cord; then tighten the leaking fitting or the large nut on the solenoid valve if the solenoid is leaking.

You can use a small screwdriver to adjust sensivity of the solar sensor. It should cause the light on the control box to come on at about 10:00 a.m. on a sunny summer day and go off around 5:00 p.m. If the light fails to come on at the right time, check operation of the solar sensor by making a short circuit between terminals 1 and 3, using a screwdriver or other tool with an insulated handle, to avoid a shock. After you short the two terminals, if the light goes on, your sensor is defective or improperly adjusted.

If you have trouble with couplers popping off, there is too much pressure somewhere in your solar panel array. Make sure that clamps are properly placed and tightened so that your panels are securely tied down. Check your top and bottom header pipes to make sure they are in a straight line, parallel with the ground. Also make sure that your feed and return pipes have not been forced into place but make a good fit, or register, with the panel headers at the coupling points.

Your solar heating system is working if the water returning to the pool is a few degrees, from 2° to 6°, warmer than the water leaving the pool. As mentioned before, a solar heater like this will take a few days to accomplish the initial warm-up, but it will transform a cold pool in January into a swimmable 70° in a typical California installation.

When your solar heater is working, the pressure gauge on your filter shows an increased pressure of up to 5 psi. And your solar panels should be cool to the touch on a warm sunny day.

Plumbing Tips for Solar-Heated Pools

Repairing a Leak

If you have a leak, the first thing is to check your clamps and make sure they are all tight. A small leak or crack is easy to repair with a patching kit provided by the manufacturer. However, if a panel is too badly damaged for simple repair, remove it and fill the space top and bottom with two 2-inch plastic pipes until the panel can be replaced.

Maintaining Suction of the Pool Sweep

Should you have a pool sweep, there are a few useful precautions. If your sweep loses suction when the solar heater first comes on, it's because air is being driven from the panels, causing the pool sweep pump to lose pressure. One cure is to locate the suction for the pool sweep ahead of the return to the panels. A second alternative is to locate the sweep saddle fitting on the bottom of a horizontal pipe. The third solution is to set your sweep timer to turn on the sweep early in the morning or late in the evening so there's no interference with the solar heating cycle.

MAKERS OF SOLAR HEATERS FOR POOLS

This is only a partial list of those specializing in solar pool heaters.

FAFCO, INC., 138 Jefferson Drive, Menlo Park, California 94025, makes the kind of black plastic solar panels described in this chapter.

Price of the Fafco panels is $2.50 per square foot, or $80 and $100 for the two sizes of panel. Fittings are $5 per panel, including tie-down straps. There is a "system pack" costing $50 which includes a vacuum relief valve, extra couplers and tie-downs, a check valve, patch kit, and instruction manual for installation and operation. Cost of the automatic control system runs $125 and the manual system is $25. According to the manufacturer, most owners of pools buy both automatic and manual controls.

This kind of solar panel, light, durable and relatively inexpensive, has other possible applications beyond heating swimming pools. A solar heater warming a large flow of water might be useful as a supplementary heater for a laundry. It could also serve in some climates to provide warmed water for a greenhouse.

FUN & FROLIC, INC., P.O. Box 277, Madison Heights, Michigan 48071, makes a panel called Solarator, 6 feet by 3 feet, formed from 190 feet of black plastic PVC tubing. Cost of a do-it-yourself system including six Solarators, pump, heat exchanger and controls is about $950.

SUHAY ENTERPRISES, 1505 East Windsor Road, Glendale, California 91205, provides only instructions and price list for materials for a homemade solar collector for pools.

SUNDU COMPANY, 3319 Keys Lane, Anaheim, California 92804, furnishes panels of ABS plastic impregnated with carbon black. Units are constructed of rectangular tubes $5/16$ inch thick and 6 inches wide placed side by side. Eight tubes interconnected with a header at each end make a panel 4 feet wide; lengths of panels are 8 feet and 10 feet. Panels are priced at $1.75 per square foot.

SUNWATER COMPANY, 1112 Pioneer Way, El Cajon, California 92020, supplies metal solar panels and complete systems for heating pools, with years of experience in so doing. Typical cost is $7 per square foot of solar panel if you install it; about $10 per square foot installed. Standard size is 8 feet by 4 feet.

9 / Backyard Barbecue Using Sunshine as Fuel

Imagine being able to cook all kinds of foods in your backyard in an ingenious sun-powered barbecue. No charcoal. No lighter fluid. You can be the first in your neighborhood to say, "I'm cooking with sunlight. Producing no smoke. And it doesn't cost us a cent for fuel."

Until recently, because charcoal and fluid to light it were relatively cheap, there was little interest in the United States in solar cookers. Only a few scientists working at universities developed clever designs for utilizing sunshine for boiling and baking foods. However, in many underdeveloped countries, where fuel is scarce and electricity is only in major cities, the need for inexpensive methods of cooking is pronounced. Several thousand solar cookers are already in use in Egypt and other Middle Eastern nations, as well as in India, Pakistan and Mexico.

These developments, including designs created by American solar scientists, have excited the interest of the Nutrition Division of the Food and Agriculture Organization of the United Nations. Some of the pioneering work in developing a plastic solar cooker at the University of Wisconsin was sponsored by a grant from the Rockefeller Foundation.

As prices for charcoal, lighter fluid and conventional barbecue grills and ovens continue to climb, it's quite possible several American manufacturers will start building solar cookers. A price tag ranging from $20 to $50, depending on the type, seems reasonable. This chapter will review the various designs of solar cookers and explain how to build a very simple one with ease.

HOW TO BUILD A SOLAR COOKER

The Telkes Solar Oven

Some years ago, just to prove that you don't have to be in the desert in midsummer to use sunlight for cooking, a lady served a group of guests a solar lunch in November in New York City. The lady was Dr. Maria Telkes, noted research scientist then working with solar energy at New York University. She put a shiny aluminum box on a wood table on a terrace and proceeded to cook a tasty meal. (See Figure 66.)

Figure 66. Highly efficient Telkes oven is easy to make yourself.

This box is one kind of solar cooker you could build for yourself if you're handy with simple tools:
1. You start by making an aluminum enclosure, with the ends open, 12 inches on each side.
2. Inside your box you provide a liner 1-inch thick of black insulating material such as styrofoam or corkboard.
3. At the top end of the cooker you insert a glass window consisting of two panes, each 10 inches square, and separated by the grooves which hold them, by an air space of ¾ inch. This forms the kind of heat trap, discussed in Chapter 3, used on solar panels. It lets in sunlight but reduces losses from infrared radiation at the top of your cooker.

4. Surrounding this window, at the top of the box, are four hinged pieces of aluminum, one on each side. These open up like the cover of a shipping carton. The underside of each aluminum flap must be kept bright and shiny, since its function is to be a solar reflector and concentrate more heat inside your cooker. Coating polished aluminum with a transparent protective plastic, such as a thin coating of epoxy, will do the job.

5. At the bottom of your solar oven, put another sheet of heavy gauge aluminum painted black, as a collector. This plate and the black insulation on the walls will absorb the sun's heat and get hotter and hotter. In fact, this kind of oven reaches 350° F in a northern latitude like New York or Chicago on a sunny day in November. In Los Angeles or Phoenix in July, it will go as high as 450°.

6. Mount the cooking chamber, containing foods, *under* the black bottom plate of the solar heat collector. A convenient way to do this is to add an insulated tray about 8 inches deep below the black plate. Provide a hinged door at the side. Then you can slide a separate, sectional tray of the food to be cooked into your solar cooker, something like a TV tray. Fasten the door of the lower "food cell," point your cooker at the sun, and soon you'll have a well-cooked meal. You can achieve whatever angle you need by simply propping up one side of the cooker with a non-flammable object, like a brick; or by building an adjustable cradle suspension.

This Telkes oven, with its four slanting reflectors of bright aluminum reflecting sunlight down through the window to cook your food, has an effective area for collecting solar energy of more than 6 square feet.

Improving the Telkes Oven

There are two improvements which can be added for making this oven perform better and store heat longer. One is to provide a space under the black collector plate, between it and the cooking chamber. You fill this space, which needs to be only about one inch deep, with Glauber's salt or sodium hydroxide in a tight container of any heat-resistant material. This salt is available at any chemical supply house, and an adequate supply will cost you about $2. It serves as a heat-storing agent. Also, this chemical helps build up heat at a faster rate and allows the oven to hold heat for an hour or so after the sun goes down.

Using the same hydroxide material in the air space between the

167

two panes of the glass window will also help. If you don't use the chemical there, put an insulating pad over the window to retain the heat when the sunlight is gone.

Cooking Efficiency of the Telkes Oven

How well does this kind of solar oven work? At her demonstration in November in New York, Dr. Telkes let her guests enjoy the aroma of hamburgers and onions without any contamination from charcoal. Out of her cooking chamber she pulled a big iron skillet with nine sizzling hamburger patties. Then she brought out two steaming aluminum double boilers, one with mashed potatoes and the other with mixed vegetables.

Guests murmured unbelievingly as they enjoyed this buffet lunch, everything "cooked better than Schrafft's" according to one compliment. The impressive fact was that on an autumn day when the outside temperature was a mild 70°, but the sun was fairly low on the horizon, it was possible wih this simple solar cooker to generate a cooking temperature of 350° and then retain the heat.

Without air pollution, cooking by sunpower is about as efficient as your conventional barbecue. In the specific solar cooker described, you can cook hamburgers in 20 to 35 minutes depending on their size and thickness. Potatoes and vegetables are ready to eat in the same time.

According to Dr. Telkes, "We have made beef stew and lamb stew. We've cooked some very delicious roasts, and we've even broiled chicken in this model. Most roasts need only about 325° or 350°, and a medium [1-pound] roast of beef takes a half hour or so. A pork shoulder takes a little longer, as is true for any boned or rolled roasts. We have cooked all kinds of fish, and of course soup is no problem at all. We have baked cakes, too. Great success with everything. No complaints from anyone about flavor, or raw spots."

Since Dr. Telkes is not only a distinguished solar physicist but also an excellent cook, her work with solar cookers has been sponsored by the Ford Foundation among others. One result is an improved model with eight flat, hinged reflectors instead of four. As a result there is a far greater surface area concentrating the sun's heat on the window of the oven. Even in a northern latitude such as Boston, this type of solar oven reaches temperatures of 400° F under a wintry sun. It will reach nearly 500° F in the southeastern or southwestern United States during the summer months, when most people use their backyard barbecues.

Simplified Indian Solar Oven

For anyone who wanted to build the simplest kind of solar oven, he could follow a design used in India.

In this case, you could build a hot box of any material, including wood, so long as you can paint the inside surfaces black. The cover of the box contains the double-paned window previously described. Below the collector plate at the bottom of the box is your food chamber.

Cooking Efficiency of Indian Oven

This extremely simple box must be kept facing the sun. It won't work at all, according to M. L. Ghai, an Indian solar scientist, unless the sun is at least 65° above the horizon. That means your use of such an elementary solar oven is limited to a few hours daily in the summer, and still fewer in winter, unless you keep tilting the window of the box so that sunlight falls directly on the black collector plate at the bottom.

Also, this box will heat up much more slowly and reach an upper temperature of only about 212° F, the boiling point of water. So you are far better off, if you want to cook with sunlight, to take the additional trouble to equip your solar oven with hinged reflectors.

ADDITIONAL TYPES OF SOLAR COOKERS

Development of Umbrella-type Ovens

The improved design developed by Dr. Telkes, having eight hinged reflectors to heat a solar oven, is similar to a whole family of cookers. These use the principle that if you reflect solar energy from a highly polished parabolic surface—a concave mirror—you will find an extremely hot spot at the focal point of the parabola.

To achieve this, you can build an umbrella with the inside coated with aluminized plastic such as Mylar or Tedlar. Then you provide a grill of aluminum mesh for a cooking pot, mounting this grill horizontally on the handle of your upside-down reflecting "umbrella."

Using a textbook on solid geometry, you can readily calculate exactly where the focal point of your particular parabola will be. This is where you place the coarse screen platform on which sits your aluminum boiler or other cooking pot. The concentrated sunlight heats the bottom of your pot.

Tarcici Umbrella-Type Oven Designs

One of the first scientists to develop a practical solar cooker of this type was Dr. Adam Tarcici of Beirut, Lebanon. While he was studying for his doctorate at the Sorbonne in Paris more than thirty years ago, he decided that inexpensive cooking by sunlight would be a boon to millions of people in Asia and Africa. Since designing his first crude model, successfully demonstrated in Nice, France, in 1949, Dr. Tarcici has developed almost thirty different models of solar cookers.

One of them has been made in quantity by the Delta Mill Company, Cairo, Egypt. This is a compact unit with a collapsible reflector of the parabolic umbrella type. Because it can be disassembled and assembled quickly, it is portable and easy to clean. You could use this kind of solar cooker in your backyard and then stow it away in the space of a deck chair—all for far less than the cost of building a brick or stone barbecue oven. And there are never any ashes to clean.

Indian Umbrella-Type Ovens

A typical solar cooker developed in India and mass-produced for a price equivalent to about $15 also utilizes a parabolic reflector made of spun aluminum. Designed at the National Physical Laboratory, New Delhi, this solar collector is mounted on a substantial steel frame. Typical dimensions of the focusing reflector are 4 feet in diameter with the cooking pot mounted on a wire grill about 2 feet above the center of the parabola. With the sun's rays focused on a pot 8 inches in diameter, solar energy is equivalent to a 500-watt electric hot plate. It will bring a quart of water to a boil in 15 minutes.

Another Indian design consists of aluminum parabolic sections mounted on steel cradles in such a way that the whole assembly may be rotated to follow the sun's movement each day. This is desirable for maximum efficiency because the greatest heat is obtained if the sun is shining directly at the center of the parabolic section.

Wisconsin Umbrella-Type Oven

Still another design was created in the United States by the University of Wisconsin, aided by a grant from the Rockefeller Foundation. One objective was to develop an inexpensive solar cooker, suitable for mass production with cheap materials, which could be used extensively in those parts of Mexico and Latin America where there is no electricity and any kind of fuel is scarce, as for example in desert and mountainous locations.

A collapsible parabolic reflector was developed by two noted solar scientists, Dr. George O. G. Löf and Dale A. Fester, both of Denver, Colorado.

The reflector has a framework like a standard umbrella frame covered with aluminized Mylar plastic sheeting laminated to durable rayon cloth. Material for the sixteen segments is die cut from a roll according to a pattern, assembled by machine stitching, and mounted on an umbrella frame having sixteen 2-foot ribs.

The material segments are cut to a special shape so that the assembled "umbrella" will have a suitable parabolic shape. The final shape of this unit depends entirely on how the material is cut, because the ribs are of spring steel and will flex as constrained by the fabric.

When it is opened up, this reflector looks like an umbrella with a highly reflective coating. In fact it's been given the name of Umbroiler. Allowing for the slight amount of shading from the sun as a result of the grill mounted on the umbrella handle at the focus of the reflector, there is an effective reflecting area of 11 square feet.

"The shape of the umbrella-type reflector is not paraboloidal in nature and does not have a compound curvature," explain Löf and Fester. "Instead it is curved only in the radial direction, from the umbrella center. Eight small portions of parabolic cylinders with wedge-shaped sides are formed. The effect of this departure from a true paraboloid is to diffuse the focus from a very small, intensely hot spot to a larger and less intensely, but more uniformly heated area. This proves to be advantageous for broiling since meat can be uniformly cooked instead of being burned in one spot and left raw in another. Also burned food in pan bottoms is eliminated."

As shown in Figure 67, the solar cooker stand is a simple folding tripod of ⅝-inch aluminum tubing hinged from a central aluminum casting at the apex of the tripod. A hole in this casting at its top accepts a metal peg attached below the main umbrella. This peg holds the umbrella on the tripod, fastened in place by a thumbscrew.

Atop the umbrella "handle" is a 9-inch by 9-inch wire grill which serves as a cooking surface. This grill is held at the correct position in the hot zone by a thumbscrew fitting attached to the ½-inch aluminum rod, the reflector shaft or handle.

Reflector altitude can be adjusted by another thumbscrew permitting you to move the umbrella up or down on the center rod.

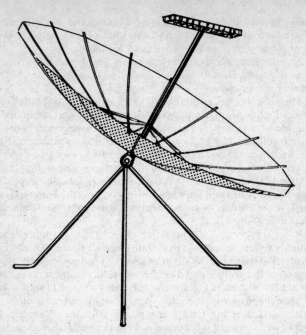

Figure 67. Assembled Umbroiler

The easiest way to change the horizontal position of the reflector is to turn the tripod stand. Grill tilt may be varied by a grill adjustment screw. If you're cooking food in pots and pans, of course, the grill must be practically horizontal. If you are broiling meat, however, you'll want to tilt the grill to focus more solar energy on it.

There's a small white cloth at the center of the reflector of the Umbroiler to give you a target for focusing. Your cooker is correctly aligned when the shadow of the grill falls on the center of this white cloth. Now you're getting solar energy reflected on the grill from the entire inside surface of the umbrella.

Cooking efficiency of the umbroiler

Tests made by Löf and Fester in Denver and Gunnar Pleijel in Sweden indicated a solar efficiency in heating water of 25%. Using an aluminum tea kettle with its bottom painted dull black, it took about 20 minutes to boil water. Imagine how useful a folding cooker like this would be for a skiing party! No matches, no firewood. Cook yourself a hearty lunch while the sun's shining.

Results with the Umbroiler showed it took 5 minutes to heat a frying pan with blackened bottom and then another 5 minutes to fry eggs and sausages. Hot dogs were cooked on the grill in 10 minutes, hamburgers in 10 to 15 minutes—the same time for grilled trout—while a thick steak took 15 to 20 minutes, depending on preference for medium rare or well done.

Possibility for mass production

Quite a number of Umbroilers were manufactured several years ago to sell for a retail price of $30. This included manufacturing cost, overhead, selling expense, and profit. About two-thirds of the manufacturing cost was for materials, the other one-third for labor.

According to the inventors, a number of minor design changes in selection of materials and manufacturing processes would reduce the costs considerably. Also, expense for materials would be lower if such a solar cooker were made in volume.

Certainly for travelers in various parts of the world where fuel is scarce, for skiers and mountaineers, as well as backyard barbecue enthusiasts, a solar cooker like the Umbroiler appears to be a useful gadget. When you see one advertised on television, you'll know some daring enterpreneur is about to make his or her fortune from cooking with sunlight.

Portuguese Cylindro-Parabolic Solar Oven

Considerable research was done by Salgado Prato, assistant engineer in the Laboratorio Nacional de Engenharia Civil, Lisbon, Portugal, in designing two models of a cylindro-parabolic solar oven. Such an oven was big enough to permit boiling about 1¼ gallons of water in a couple of hours, roasting a 5-pound chicken, frying fish, eggs and cornmeal.

In this cooker, the solar energy is collected by a parabolo-cylindrical reflector of polished aluminum and directed to a narrow glass window in the lower part of an insulated cylinder with a horizontal axis. The window is at the focus of the paraboloid. The food to be cooked is placed in the oven just above the window.

According to Prato, this design makes it possible to increase the solar concentration and reduce heat losses as compared with other oven-type sun cookers.

Stam Solar Oven and Heat Accumulator

Another experimenter is H. Stam in Amsterdam. He believes that it's important to make cheap solar kitchens for many parts of the world, particularly in desert or semi-desert regions of Asia and Africa where use of brush and dung for household cooking takes from those parched lands the small amount of fodder to feed animals, and also destroys the only indigenous fertilizer.

His proposal is to use fairly large but inexpensive reflectors. These can be used for cooking for a family or a small tribe during the day as well as heating domestic water. When it is not in use for these purposes, the reflector can be heating an accumulator containing some cheap eutectic material like crystallized magnesium chloride, which melts at about 245° F. This is melted into liquid form during the day by focused sunlight from a parabolic reflector faced with aluminum foil. Then at night, as the eutectic material becomes a solid again, it gives up its heat and permits cooking after the sun goes down.

Solar Steam Pressure Cooker

If you want a useful solar cooker heated by superhot water from a collector panel, William B. Edmondson, publisher of *Solar Energy Digest* and developer of the highly efficient SolarSan collector panel, has built several models. Here is a description of his latest and best solar steam pressure cooker—another of his inventions on which a patent application has been filed.

According to the inventor and his wife, this cooker has been used for months to cook beans, potatoes, eggs, meat and vegetables. Figure 68 shows a cross-section of this unit.

You start with an ordinary stainless steel pressure cooker and surround it with a stainless steel steam jacket. If you're not an experienced welder, you can find a welding shop to build this part of it. There are three openings in the steam jacket. The one at the right in the drawing is an inlet for superheated water (water heated above its boiling point) and steam from the solar panel. The bottom opening returns condensed steam and water to the collector panel. This third, shown at the left, is a try cock to let out air and excess water when you start using the cooker.

Note that the welded outer jacket contains a layer of high temperature insulation between this outer shell and the pressure cooker itself. Edmondson has found it best to build up thin layers of glass wool alternating with aluminum foil that's reflective on both sides. Quoting the inventor: "Insulating the cooker is proba-

174

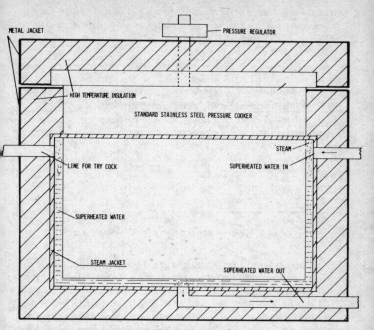

Figure 68. Cross-section of Edmondson's solar steam pressure cooker.

bly the most important thing you can do to insure that it will cook successfully, so don't skimp on the insulation. Make it as much like a vacuum bottle as possible. If you have a high vacuum pump handy, you might even pull a high vacuum between the outer jacket and the cooker itself, if you have used heavy enough metal to withstand the vacuum. The better the insulation, the smaller your SolarSan (collector panel) can be."

Based on the experience of the Edmondsons, a single collector panel of the SolarSan type, described in Chapter 3, will give you enough superheated water by 10:30 a.m. on a bright sunny day so that the cooker is ready to hum. That's if you've done a good job with the insulation and your solar panel is efficient.

If you want to have this solar cooker inside your house, it's possible to do it under certain conditions. Keep the lengths of your inlet and outlet tubing to a minumum, not over a few feet long. It will help to insulate these lines. Remember that you're using superheated water from the solar panel. Tests with the Edmondson SolarSan on a day in January when the flow rate of water was set at a fairly low rate, namely 0.234 pounds per minute, show the following results when the water was left running from 8:00 a.m. to 3:00 p.m. (see Table 3).

Time	Ambient Temperature	Inlet Temperature (Water Supply)	Outlet Temperature	Temperature Rise
0900	52°	58°	140°	82°
1000	65	63	185	122
1100	75	71	218	147
1200	80	74	235	161
1300	82	75	236	161
1400	85	74	225	151
1500	85	74	195	121

Table 3. Heating Performance of SolarSan Panel

Evidently it would be desirable to do your cooking, under the conditions shown in this table, between 11:00 a.m. and 2:00 p.m. For convenience, you may want to weld some tabs on the bottom of the outside jacket of the cooker so that you can attach this part of the unit to a flat surface. By so doing you can remove the pressure cooker section when you want to empty it or clean it, without having to disconnect the tubing running from the solar panel.

Edmondson and his wife have found a still better way. They put a cup of water in the pressure cooker, and place the small food elevator that comes with the cooker at the bottom. Then they put another open pot, containing the food to be cooked, on the elevator. Thus there's no problem with cleaning out the cooker. "All you have have to do is wipe it out occasionally with a soapy cloth and oil the rubber gasket," says the inventor.

Assuming that you are dedicating one solar collector panel to supply superhot water for cooking, you'll need to provide a valve at the lowest portion of the panel for filling this with water or an antifreeze solution. The easiest way is to connect your garden hose to this valve in the panel, then let the water flow through the whole system, including panel and cooker, until water starts coming out of the try cock. Then turn off this filling valve tightly and leave the try cock open until the solar panel builds up steam. At this point, steam pressure will blow excess water and steam out of the system. When all the air is out of the system and the water level is just below the try cock, shut off this exhaust valve and start cooking.

176

If all your plumbing connections are tight, you have built a "hot pipe" which should be liquid-tight for many months, possibly years, if you've filled it with an antifreeze solution. Because you have a closed system, when the temperature goes down at night the liquid will be under a vacuum. This makes it possible for the water or antifreeze solution to heat up rapidly when the sun comes up. By 10:30 or 11:00 on any bright morning your cooker is perking.

Because you want the perking action caused by thermosiphoning, be sure to put your cooker about a foot higher than the top of your solar panel, so that the hot water will naturally rise to it. For a backyard installation, this should present no great problem. If you want your solar cooker indoors, this may take some ingenuity. Keep in mind, however, that you don't have to mount your collector panel on a roof. In fact, you may be able to place one of the new Corning Glass solar panels on a south-facing wall. This panel, described in Chapter 3, delivers liquids at temperatures up to 300° F or more. Such a solar cooker would cut gas or electric bills (depending on your type of stove) even further.

10 / Using Electricity from Your Own Solar Power Plant

In every previous chapter there's been emphasis on the fact that you can benefit from practical applications of solar energy *now*. This is certainly true in regard to heating and cooling your home, heating your water supply, your swimming pool, and cooking your food. But building a small solar power plant, large enough to provide from 5 to 10 kilowatt-hours (kwh) of electricity for your home, seems to be a year or more away from an economical reality. However, there are several techniques that look promising indeed, and we will now take a look at them.

APPROACHES TO GENERATING SOLAR POWER

Considering only a small solar power plant—there's a discussion of ways to build large generating stations using solar energy in Chapter 12—the most interesting techniques can be rated in terms of immediacy: (1) Heat from solar collectors is used to operate turbogenerators, with modified Rankine cycle heat pumps driving the generators, and possibly the Banks nitinol engine. (2) Solar cells provide d-c power, which is stored in batteries and converted to a-c for home use. (3) Solar furnaces heat thermoelectric materials or thermionic converters to generate enough kwh for your home.

In each case it seems likely that the electric power generated by solar energy will be stored in batteries. In Chapter 11 are some comments as to the massive research being done to develop lighter, more efficient storage batteries at a reasonable price—so that perhaps your 198? car, for commuting purposes at least, will be an electric model.

MANUFACTURERS' PLANS FOR SOLAR POWER PLANTS YOU CAN ASSEMBLE YOURSELF

Omnium-G Parabolic Aluminum Collector

Several companies are working on small solar power plants which you may be able to assemble yourself. One example is a firm called Omnium-G Enterprises in Anaheim, California. Housed in an industrial tract building, this venture is the dream-child of its four founders. All are engineers, moonlighting to create a solar power plant capable of supplying electricity and hot water or hot air for most of the needs of a typical family in a six-room house.

Block diagram of system

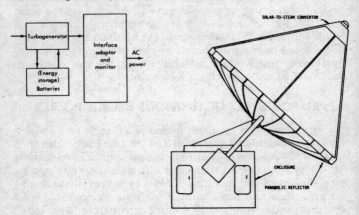

Figure 69. Omnium-G design for a backyard solar power plant. (See page 000 for updated information.)

Shown in Figure 69 in block diagram is their method of producing electric power from sunlight. This method involves the use of a parabolic reflector, shown in Figure 70a. This highly reflective, curved device concentrates solar heat at a focal point due to its particular geometric shape. A large parabolic section like the Omnium-G reflector—about 13 feet on a side—generates heat at its focus equivalent to approximately 2½ million times the solar heat collected on a square inch of flat solar collector panel. Figure 70b is a view of the parabolic aluminum collector with the solar furnace, or concentrated heat exchanger, located at the focus of the

Figure 70a. Parabolic reflector designed by Omnium-G.

Figure 70b. Early parabolic reflector with heat exchanger at focal point. (See Figure 69 for current design.)

parabolic section. Solar heat turns the water in this small stainless steel boiler into steam, which drives a turbogenerator estimated to produce enough electricity for an average household.

There have been quite a number of earlier solar generators based on this principle. Among them are systems described by Dr. Harry Tabor, director of the National Physical Laboratory of Israel, Jerusalem, and Professor Valentin Baum, head of the Heliolaboratory, Moscow, at the United Nations Conference in Rome in 1961; as well as systems by numerous other inventors in the United States and abroad.

Why, then, should Omnium-G engineers work nights and weekends on an old concept? Because certain components of their conversion system appear to be patentable. Further, they believe their little power plant, when manufactured in some quantity and made available to mechanically inclined persons in kit form for backyard assembly, will be available at an attractively low price, around $2,000.

Basic Construction of Omnium-G Power Plant

Here's how this solar energy conversion system works. A polished aluminum collector is mounted on a cradle having an automatic control unit which always points the center of this parabolic section directly at the sun. This is one useful feature. A small cylinder containing an electronic photodetector is the trigger for a servo control circuit operating an electric motor. Once the detector is pointed at the sun, it's as faithful as a bird dog pointing at a covey of quail in a bush. The motor turns the reflective collector in elevation and azimuth (vertically and horizontally), so that its center is always aimed at the sun to collect the greatest amount of heat possible.

Mounted at the focal point of the collector is a small cylindrical heat exchanger. This unit is considered by Omnium-G to be patentable, and patent applications have been filed. Made of materials which will withstand intense heat, this exchanger is truly a solar furnace. The heat of the sun is magnified at the focal point of such a large collector, which is about 13 feet by 13 feet.

Thus, water brought into the heat exchanger, or Steam Cycle, (see Figure 69) is superheated in a small space under pressure. The output from the exchanger is high-temperature steam which drives a small turbogenerator. From this generator the a-c power produced goes to an interface panel for distribution to the home, and part of the electricity is used during daylight hours to charge storage batteries. This bank of batteries provides the reservoir of electrical energy for use at night and during cloudy/rainy/snowy days.

Various Models of the Omnium-G Power Plant

An interesting feature of Omnium-G's heat exchanger is that it is available in various models. One produces only steam to drive the turbine. A second type furnishes somewhat less steam, but has an outlet for hot water which can be used for space heating and/or to supplement a conventional water heater. The third type of heat exchanger produces steam to drive the turbogenerator, and has an outlet for hot air fed into the home's forced-air heating system. In this latter case, there would be intermediate storage of the hot air by means of a rock bin as described in Chapter 4.

According to the firm's calculations, the efficiency of their system approaches 50% if the only end product is electricity. That is, in the Los Angeles area, the sun provides an average energy input to the parabolic collector amounting to about 16 kwh. If the heat exchanger produces only steam to drive the turbogenerator, there is an electrical output to your home of 8 kwh. Should you need less electricity and want hot water, the exchanger is more efficient and produces 3 kwh of a-c power as well as 7.5 kwh equivalent energy in hot water.

Another interesting concept for storing solar energy for later conversion into electricity is to use the Omnium-G heat exchanger to pump air, under pressure, into conventional cylinders. The compressed air obtained in this manner may later be released to drive the turbogenerator and thus obtain electric power.

Efficiency of Omnium-G Power Plant

Considering this mini-power plant on an annual basis, it is projected to deliver 43 megawatts (Mw) per year if installed in a backyard in the Los Angeles area. This is 8 Mw more than the average household needs to take care of all its electrical requirements as well as its needs for natural gas. The typical family's gas and electric consumption over four seasons comes to an average of 35 Mw per year.

Another analysis by Omnium-G shows the effects of latitude and climate. Using the Omnium-G system in places like Houston, Rapid City, or Atlanta, you would have to depend on some supplementary source of electric power for part of your requirements. Part of this could come from conventional electric and gas mains. Quite possibly most of the necessary energy for space heating and hot water would come from collector panels of the various types previously discussed, while your solar power plant and its associated storage batteries furnish all the electricity you need.

Certainly in such sunny climates as Phoenix, Tucson, Los Angeles, San Diego and similar cities all over the world, a system of the type being developed by Omnium-G will provide most of the domestic energy requirements. In Los Angeles, more energy is generated on clear, dry days, far less during humid smoggy weather. Similar situations occur in such "ideal" sunshine areas as El Paso, Las Vegas, and almost any cozy oasis in the Sahara. (And if you think that's funny, it's because you don't know of all the buildings heated and cooled with solar energy in various countries in North Africa.)

Projected Cost of Omnium-G Solar Kit

Anyplace where there is a total of 240 days or more of annual sunshine, say the engineers of Omnium-G, you'll be able to purchase their components in kit form and build yourself a practical household power plant at reasonable cost.

Just what is reasonable cost? These inventors are basing future prices on how many systems are manufactured. Their marketing plan is aimed in large measure at the export market—areas of Asia, Africa, and South America which are remote from electric power.

This seems like a practical idea. In the book *Direct Use of the Sun's Energy* by Dr. Farrington Daniels, the author points out: ". . . there are two billion people in the world without electricity and it will be a long time before they can be served. Moreover, in communities of small rural users widely distributed, the cost of distribution of conventional electric power will continue to be so great that solar energy may perhaps be competitive."

This was written in 1964, long before utility rates soared as a result of higher prices for fossil fuels. Costs of oil, gas and coal appear certain to climb still higher. Thus, even the utility companies are taking a keen interest in solar energy.

According to Dr. Daniels, there are specific needs for small solar power plants, in addition to the market among rural and remote homesteads, farms and ranches. In India, China and many other countries with literally millions of small villages, there is a need for solar-generated electricity to increase the productivity of so-called village industries—weaving, pottery-making, leather goods manufacture. In Pakistan it would help the farmers greatly if there were some twenty thousand solar-powered pumps to deliver fresh water into irrigated areas and thus to rinse out salt deposits which make the land useless for farming.

Even in the United States, there are about thirty-one thousand homes which are not served by public utilities because they are too remote. Another sixty-two thousand households indicated in a random, cross-country survey that they would like to have solar power plants as back-up units when available at reasonable cost.

Based on their mid-1975 projections, the founders of Omnium-G think they should sell two thousand five hundred power systems in their first two years of full production. This would mean the calendar years 1976 and 1977, if all goes well. If they reach a level of about two thousand units a year, they expect the price to come down from about $7,000 to less than $4,000 per system. This figure might be still lower for do-it-yourself persons who buy the components in kit form and assemble them. You might be able to build your own backyard power plant for under $3,000 in a couple of years with the Omnium-G approach.

If you total your electric and gas bills for the past few years, assuming you live in a house with six or more rooms, you'll find that a solar power plant should pay for itself fairly rapidly. Once the initial investment is amortized, maintenance costs are low. You'll only have to replace battery modules perhaps once every five years.

International Research Associates' Parabolic Collector

Many other companies, large and small, are working on similar small solar power plants. One small firm in northern California is International Research Associates (IRA), Livermore.

According to Dr. Henry Cheung, noted physicist and president of IRA, his company's system should be still more efficient than the Omnium-G design. It utilizes a considerably smaller parabolic section which is cheaper and easier to control. To make up for its smaller collection of solar energy, the IRA heat exchanger utilizes a liquid with a much higher molecular weight than water, such as freon. This means that a more efficient turbogenerator can be utilized.

On the other hand, the overall system is somewhat more complex because the freon must be recycled through a condenser. So far it's hard to determine whether there are cost advantages in favor of a system like IRA's, but because of the smaller size of the parabolic reflectors it should be easier to install in your backyard.

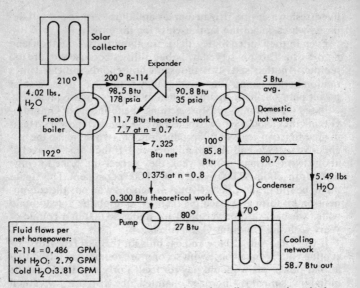

Figure 71. Power generating system using solar collector panels and a freon boiler developed by Kinetics Corporation.

Kinetics Generator with Flat-Plate Collectors

Another system for generating electricity for solar energy has been announced by Kinetics Corporation, Sarasota, Florida. A block diagram of the operation of this small solar power plant appears in Figure 71. President of the company is Wallace L. Minto, who previously developed a pollution-free automobile engine operated on a freon cycle.

Basic Construction of Kinetics Generator

In this solar power generator, R-114 is a freon compound. Minto uses a typical array of flat-plate solar collector panels. These can be any of the liquid-type units described in previous chapters, just so long as water in the panels is heated to at least 195° F. The heated water, preferably at a temperature of 210° F or more—which makes designs like Corning, SAV, Piper, SolarSan, Garden Way, Sunway, Revere, Reynolds, PPG and a few others look quite adequate—passes through the freon boiler, converting the freon into gas. The gas enters an engine where it then expands and operates a turbogenerator.

186

Figure 72. Demonstration model of Kinetics generator lights one hundred 40-watt bulbs.

Efficiency of Kinetics Generator

In a demonstration system, this generator produces enough a-c power to light a panel of one hundred light bulbs, 40 watts each. (See Figure 72.) Minto believes this 4-kw generator produces enough power for the average home. Note in Figure 71 that this power plant also produces domestic hot water at a temperature of 100° F through a heat exchanger before the gas is cooled by a condenser to return to its liquid state.

One economical feature of this Kinetics system is that the piping grid for the cooling network is laid a few inches below grass root level, according to Minto. By so doing, the system furnishes low-temperature water for the condenser, no matter how hot the ambient air temperature may be—and it can be very hot in Florida.

This method of cooling the condensing water eliminates the expense of a cooling tower, provides colder water, and thus helps to achieve a more efficient Rankine cycle.

An interesting sidelight is that the heritage of Minto's freon engine is a development financed by Nissan Motor Company, of Japan. This company supplied $1 million to Kinetics for designing a pollution-free Rankine-type engine mounted in a Datsun II automobile. Negotiations are now under way looking toward possible mass production of this automobile engine.

Cost of Kinetics Generator

Meanwhile, Minto and associates have set up Sun Power Systems in Sarasota to market a 10-kw commercial solar power package. This includes a freon engine, vaporizer, condenser, pumps and controls. "The solar collection panels, storage tank and other components will be supplied by approved manufacturers that have met the performance efficiency requirements of Sun Power Systems," says a company official.

It is expected that initial installations will be in commercial and institutional buildings where this solar energy system will serve to provide most of the electricity and hot water required. By furnishing energy during daytime hours when the electric utility's load is peaked, the Minto system can be useful to the power utility industry—as well as saving money for the owners of such solar installations.

In his commercial design, Minto achieves 10 kw of electric power when the solar-heated water is at 160° F. Water from collector panels at 140° F yields 5 kw. When the water is at 180°, the output is 15 kw. Two firms in Florida have purchased this commercial package at a current price of about $10,000 per installation.

It would appear that by adding storage batteries to the system diagrammed in Figure 71 so that surplus power is used to charge the batteries, it should be possible to have a small solar power plant that will have a few days of carryover capacity during bad weather. Storage of surplus solar energy, either in a bank of batteries or in a large hot water tank or rock storage bin suitably insulated, seems highly desirable because the Minto system delivers increased amounts of electric power as the water or antifreeze mixture in the solar panels gets hotter during the day. By using Corning collectors, for example, it should be possible to go quite rapidly from an output of 5 kw to 20 kw on a clear sunny day.

Under way as a new project of Sun Power Systems is development of a 50-kw system package for remote locations where fuel and electric power are not readily available, and for reducing utility bills in installations such as small shopping centers, schools and office or apartment buildings.

Solar Generator Using Rankine Cycle Heat Pump

A new Rankine cycle heat pump which can also generate electric power and which is driven by solar-heated fluid is under development by a team in Ohio under the direction of Dr. James A.

Eibling, manager of the solar energy systems programs at Battelle Columbus Laboratories. This unit was described in a recent issue of *ASHRAE Journal*.

In this Rankine pump, both expander and compressor include rotary vanes of a two-piece design such that the tips are free to float and pivot against the inner wall of the housing. The vane tips generate a lubricating film, and the pump needs no liquid lubricants. Financed by a National Science Foundation grant of about $90,000, this Battelle system can function as a heat exchanger which can be used for cooling a building in summer and heating it in cold weather. When neither cooling nor heating is required and the solar energy in the working fluid is adequate, this Rankine pump will drive a motor generator and produce about 10 kw of electric power.

A useful feature of this design is that, when no solar energy is available, the generator can be switched around to become an electric motor driven by utility power and thus furnish cooling or heating during standby periods. This means that a unit of this Battelle design with suitable control circuitry could function all year for home air conditioning, to furnish part of a home's electric power, and store surplus energy in batteries or other means of storage. Contracts by manufacturers interested in producing this unit with a license from Battelle are being negotiated. Therefore, installations for your house or commercial building—and price tags for the units—are perhaps a year away.

Roesel Generator for Electricity, Heating and Cooling

Another novel electric power generator which can use solar heat from collector panels, or preferably from a small solar furnace such as the Omnium-G or IRA designs, is the unit designed by John Roesel and described by E. F. Lindsley in *Popular Science,* October 1974. It is a modified Stirling engine using such liquids as Monsanto Therminol and Dow Chemical Dowtherm which remain liquid over a wide temperature range. It includes a highly efficient magnetic pump. This unit can be used for heating or cooling a building as well as for driving a motor generator and producing 60-Hz electric power from the output flywheel. Solar heat is applied to an inert pressurized gas such as argon or helium to move the liquid and operate the hydraulic motor.

Future Nitinol Heat Engine

Much more remote at present from a practical solar power generator is a heat engine developed by Ridgeway Banks of the Lawrence Berkeley Laboratory of the University of California. He uses loops of Nitinol, a nickel-titanium alloy which changes shape as it is heated. That is, the Nitinol loop tries to remember the shape it had when it was being annealed. Banks has put such loops in a bath containing cold water on one side, and a cylindrical container holding hot water, heated by a solar collector panel, on the other side.

The Nitinol loops are mounted on drive rods. These wire loops are limp while they pass through the cold water. But when they are moved up a ramp and dropped into the hot water, they "remember" their original shape and try to straighten out. The inner end of the loop is fixed to the drive rod, while the other end is free to slide and push against the outer ring of a wheel mounted above the cylindrical bath. Pushing action of a number of loops makes the wheel turn, moving a shaft.

This could drive a motor generator. After many millions of rotations, and therefore flexures of the Nitinol loops, this unusual alloy shows no signs of wear or fatigue. The inventor sees future possibilities in creating an engine of this kind using sun-heated water from collector panels. Or, on a much larger scale, using the differences in temperature of seawater from a tropical surface at perhaps 75° F to the chill of under 35° F at a depth of perhaps 2,000 feet.

UPDATED INFORMATION

Omnium-G has recently improved its design. This has achieved the advantages indicated by the revised drawing on page 180. One major improvement is that the reflector is now an inverted parabolic "flower." Each section is separated from the adjacent "petals" by about 1.5 in. This makes the solar reflector much less susceptible to wind loading, and much easier to aim at the sun automatically—via its photodetector and servo control system—when high winds blow.

Other mechanical and electronic improvements have been made, including optional use of compressed air containers ("bottles") as an energy storage system. Thus, when the sun is not shining, air compressed by solar energy drives the turbogenerator.

These improvements have made Omnium-G's solar power plant lighter in weight, easier to produce and ship. Orders will be taken in October 1976.

11 / Solar Cells and Your Future Commuter Car

As the price of gasoline keeps climbing, it's inevitable that there should be increasing interest in other kinds of cars than those with internal combustion engines. Electricity is a logical substitute, especially if this electric power can be derived from the sun's energy. Moreover, it is not necessary to build a solar power plant to do this.

An elegant way to produce electricity from the sun is the direct approach using photovoltaic devices, or solar cells. These are a group of materials, all of which are *semiconductors*—classified between the good conductors of electricity like copper, aluminum and other metals, and the insulators, or poor conductors of electric power, like glass and most ceramics. The specific characteristic of solar cells that makes them useful in converting the sun's energy is that these materials, when exposed to solar radiation, produce electrical energy.

In solar cells, heat is transformed into electric power entirely by electronic methods. This involves the excitation and massive flow of electrons in the semiconductor without any visible physical or chemical change or movement. There are no moving parts in such a photo-generator—no gears, wheels, pistons, or heat exchangers handling fluids or gases. Favorite materials used for solar cells are silicon wafers containing minute impurities; thin films of cadmium sulfide plated on copper sulfide and other metallic compounds; and gallium arsenide. There are many other possible photovoltaic substances, but these represent the most popular in terms of research and development efforts in various parts of the world.

A BRIEF HISTORY OF SOLAR CELLS

The first applications of solar cells began less than twenty-five years ago. In 1953 it was thought that the conversion efficiency of such materials was very poor indeed, about 0.6%. Quite recently, however, scientists of Communications Satellite Corporation have announced silicon cells with an efficiency of about 20%; and commercial devices with efficiencies of 12% to 14% have been available for some time.

Since solar cells have been used on panels to power satellites and space vehicles, or to supply electricity for navigation and communications equipment on oil wells far at sea or at other remote locations, the price of these photovoltaic devices has come down dramatically. Dr. Valentin A. Baum, noted Soviet solar scientist, pointed out at the United Nations energy conference in 1961 that silicon "solar-batteries" had been reduced in price from $500 per watt in 1959 to $275 per watt in 1960 and $175 a year later.

Declining Cost of Solar Cells

At present, the price of these silicon solar cells ranges from $15 to $30 per watt, depending on whether you price merely the device itself or include it as part of a system installed on a roof to deliver electric power to operate appliances in a building. Even this price is far too high to permit solar cells to compete effectively against other solar energy collectors such as the various kinds of panels described in earlier chapters. But scientists and engineers working in this field have a target: they expect to be able to supply photovoltaic power at $.50 per watt by 1980, in limited quantities of devices, and at a somewhat lower price in huge quantities by 1985.

Certainly, the whole history of semiconductor development by the electronic industry makes these targets appear not only feasible but modest; they may well be achieved somewhat sooner. There is a massive effort under way now, directed by the Energy Research and Development Administration (ERDA) with several National Aeronautics and Space Administration (NASA) facilities, including the Jet Propulsion Laboratory (JPL) of the California Institute of Technology, Pasadena; Godard Space Flight Center, Beltsville, Maryland; the National Science Foundation, Washington, D.C.; and a large team at the Sandia Laboratories of Western Electric Company, Albuquerque; as well as many other industrial research organizations.

HOW SOLAR CELLS WORK

A typical silicon solar cell has two layers, as shown on the left in Figure 73. Depending upon the tiny amount of impurities added

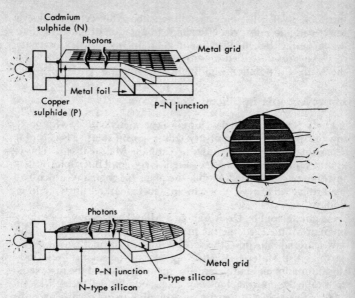

Figure 73. Components of two types of solar cells.

to the silicon, one layer is called a P-type because it conducts positive charges called holes, while the other layer is called an N-type because it conducts negative charges called electrons. Between the two layers is an intermediate surface called a P-N junction and formed by means of another diffused impurity in the silicon cell.

When an energy particle from sunlight, called a photon, strikes near the P-N junction, it produces both an electron and a hole (absence of electron, or a positive charge). As photons continue to strike the junction, electrons move toward the N-type layer and holes move toward the P-type. These charges result in creating a voltage across the cell. If you have enough silicon cells connected together, you can provide adequate electricity to operate a d-c light bulb directly from sunlight.

A cadmium sulfide (CdS) cell, diagrammed on the right in Figure 73, consists of an N-type cadmium sulfide film on the backing of P-type copper sulfide (Cu_2S). Between them is a P-N junction. Operation of this cell is similar to the silicon type.

Among the scientific teams working on devices of this kind is one at Sandia under the direction of Dr. D. G. Schueler, supervisor of the Solid State Electronics Division. Their three main objectives are: development of cadmium sulfide and copper sulfide thin film solar cells; polycrystalline silicon film cells to achieve lower cost and improved efficiency; integration of solar cells into a solar system like the installation at Solar One House, University of Delaware—described more fully later in this chapter—and on the Mitre Corporation building in Washington, D.C.

Improvements in Silicon Solar Cells

One of the leaders in developing a commercial process for making cheaper and better polycrystalline silicon solar cells is Mobil Tyco Solar Energy Corporation, Waltham, Massachusetts. This is a joint venture between Tyco, a relatively small but sophisticated instrument maker, and Mobil, the giant oil company which has bankrolled this new program to the extent of thirty million dollars.

As pointed out by Dr. Abraham I. Mlavsky, executive vice president of Mobil Tyco, the big problem is not the initial cost of the raw material. Silicon is the second most plentiful element in the earth's crust. Materials needed to make silicon solar cells, including aluminum and boron, are also plentiful. Cost of the raw materials to make one panel of solar cells, good for delivering 1 kw or 1,000 watts of electric power on a satellite in outer space, is only $150. But the cost of processing this material runs the price up enormously—at least $50,000 per kw.

A reason for this is that the conventional method for making single-crystal silicon is to slice an ingot of pure Si into thin slices. The labor cost is very large. And the cutting or grinding process destroys more than ¾ of the ingot.

So the Tyco scientists began about nine years ago to develop a method for growing Si crystals. Actually, the first material grown by their new technique was sapphire for microcircuits. They called their process "edge-defined, film-fed growth (EFG)" and have nearly one hundred patents in existence or pending.

Here's how EFG works to produce long, continuous ribbons of silicon in a method that may reduce the cost of silicon solar cells. A die is lowered into molten silicon. The hot liquid rises through its center by capillary action. The silicon flows to the top of the die and no further. Shape of the die makes the silicon crystal into a ribbon. Dr. Mlavsky of Mobil Tyco has produced ribbons up to 30 feet long and expects to reach 100 feet by the end of 1975. A typical ribbon is 1 inch wide and should be about 0.004 inches thick. This thickness is sufficient to absorb all the solar radiation that a useful photovoltaic converter needs.

A parallel effort is to make sure that these silicon ribbons have a conversion efficiency of 10% or better. This is now being achieved, but the goal is to reach something like 16% efficiency and, correspondingly, reduce the area of solar cells needed for any specific installation.

Still another thrust of the Mobil Tyco research is to grow many long ribbons at once from a single crucible of molten silicon. This is going to take time, but the experience of Tyco with sapphire makes the goal look reasonable. Sapphire tubes for arc lamps are now mass-produced by Corning Glass under license from Tyco with eighteen sapphire tubes at a time. These sapphire arc tubes for efficient sodium vapor lamps are made at lower cost than ceramic arc tubes and are hence replacing them. The raw material is a semi-precious gem and more expensive than ceramic. But the manufacturing process is more efficient.

Cost Estimates for Future Production

Mobil Tyco has estimated the cost of producing twenty silicon ribbons simultaneously with each ribbon being 2 inches wide and whatever length—perhaps 100 feet—seems economical. With the cooperation of major producers of the raw silicon, such as Dow Chemical, in reducing the cost of the silicon to perhaps $10 per pound, it seems as if manufacturing cost might come in at $15 per pound. This would mean a cost of $125 per kilowatt for silicon ribbons.

Another advantage of the ribbons is that this EFG material can be converted into solar cells, with appropriate packaging and leads for wiring, by automatic machinery. This again will result in savings.

As Dr. Mlavsky says: "Despite the enormous cost of solar cells developed for space applications, the overall process of converting sand (silicon dioxide) to solar cells is much less complex than, for example, the production of an automobile from iron ore, various other minerals, petrochemicals, etc. And the automobile sells for $2-$3 per pound at retail!"

So within a few years, after an investment estimated to be about one hundred million dollars in development and improved production machinery, we should have solar cells at a cost of $.50 or less per watt. Then this technique of generating electric power will be highly competitive with other methods. And our supplier of energy, the sun, is considered inexhaustible for at least another five billion years.

OTHER SOLAR CELL RESEARCH

Gallium Arsenide Solar Cells

One of the most important developments has been under way for several years at the Solid State Research Laboratories of Varian Associates, Palo Alto, California. This solar cell, shown in schematic cross-section in Figure 74, has an upper layer of P-type aluminum gallium arsenide (AlGaAs) exposed to sunlight. Below this is a layer of P-type gallium arsenide (GaAs), and then a relatively thick substratum of N-type gallium arsenide. As indicated in the accompanying graph plotting conversion efficiency versus thickness in microns (millionths of a meter) of the P-type aluminum gallium arsenide upper layer, the efficiency of this type of solar cell is quite high and can exceed 20%.

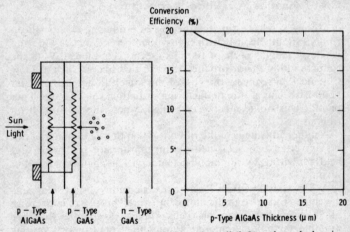

Figure 74. Construction of new Varian solar cell (left) and graph showing its efficiency in converting sunlight to electricity (right).

When news of Varian's improved gallium arsenide solar cells was announced in the public press on June 24, 1975, the price of the company's stock jumped about 15% on the New York Stock Exchange. Such is the glamour of solar energy, and the excitement in the business world about the race to manufacture solar cells which will produce electricity efficiently and economically.

According to Ronald L. Bell, director of the Varian laboratory where this research is being done, the structure of the cell permits operation at temperatures up to 360° F (200° C) or higher without providing a static cooling mechanism, such as a heat sink. The new Varian unit, shown in Figure 75, includes a concave reflector which collects sunlight and focuses it on the aluminum gallium arsenide surface of the cell. This makes it possible to increase the solar energy striking the cell by a factor of one thousand. Results of this concentration are charted in Figure 76. Note that the verti-

Figure 75. Magnified view of aluminum gallium arsenide and gallium arsenide solar cell in concentrator mount by Varian Associates. Cell's active surface is actually ⅓ inch in diameter.

cal scale for current is logarithmic and demonstrates the effect of the concentrator design in greatly increasing the power output of the cell. The triangles, marked by arrows, show the maximum power point of each curve.

Thus the Varian solar cell has an output of 10 watts with a cell diameter of only ⅓ inch. The research team doing this work is planning to connect one hundred cells in a rooftop array so that the reflectors are mounted on a rotating frame to keep concentrated sunlight focused on the cells. By using one hundred cells in series, mounted this way, it should be feasible to generate 1 kw (1,000 watts) from a relatively small solar panel.

While Ronald Bell and others of the Varian research group are enthusiastic about their results to date—see papers listed in the

197

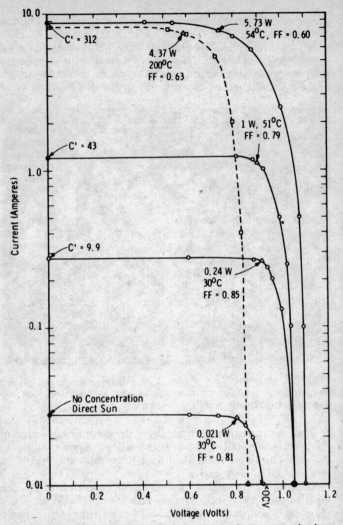

Figure 76. Effects of varying solar concentration on the current and voltage produced by the Varian gallium arsenide solar cell.

bibliography—they are not yet ready to set firm dates for their goal of producing electricity from GaAs cells at a few cents per watt. "We're some tens of millions of dollars away," Bell admits. "Even

198

so I think our approach is more certain of success than the silicon or cadmium sulfide solar cells."

Other companies doing research with gallium arsenide solar cells include IBM and RCA; these companies are reporting efficiencies as high as 18%.

Solar House Powered with Cadmium Sulfide Solar Cells

One of the most impressive experiments is being conducted at Solar House in Newark, Delaware. This program, under the direction of Dr. Karl Boer and Dr. Maria Telkes of the University of Delaware's Institute of Energy Conversion, involves one of the most advanced and ambitious experimental solar homes.

On the roof are twenty-four panels, each 8 feet long by 4 feet wide. There is an outer sheet of Abrite-coated Plexiglas, an airspace, and then a sheet of glass. After passing through this double glazing, solar radiation goes through a 1-inch space filled with dry gas, such as dry nitrogen or helium, and then hits a planar array of cadmium sulfide solar cells. Each cell has cadmium sulfide on the surface backed by a coating of copper sulfide. Fabrication of these arrays has been aided by a grant of three million dollars from Shell Oil, yet, so far, only three of the twenty-four panel spaces actually contain arrays of solar cells.

To keep the cells cool and also to gain heat, air is blown upward toward the peak of the roof by a fan. This hot air, as well as air from a vertical collector on the south wall of the house (a conventional air-type panel with black aluminum absorber), is utilized with the triple eutectic storage system designed by Dr. Telkes described in Chapter 6.

These solar cells have a 20-micron vacuum-evaporated film of cadmium sulfide (N-type) on an immersion-formed layer of P-type copper sulfide. Over the upper surface is a transparent grid electrode protected by mylar. Output of such a cell at noon on a sunny day in Delaware is between 15 and 20 milliamperes of current per square centimeter of surface at 0.37 volt. An efficiency of conversion of about 8% is hoped for although the present level is less than 5%.

At present the electric power in the house is used to operate d-c lights and heaters, with surplus stored in a 180 ampere-hour battery. Eventually, when others of the twenty-four panels are equipped with solar cells and tied into the system, there will be a d-c to a-c inverter installed to provide conventional 60-Hz power for the house. During the winter of 1973-74, it was estimated that 67% of the heating requirement for this house was furnished by

199

solar energy. In the 1974-75 winter the amount went up to 80%. Costs of the house were partially paid for by eight electric power companies. Throughout the country, utility firms, as well as manufacturers, are taking a keen interest in ways whereby they may fit their activities into the future of solar energy.

IMPROVEMENTS IN STORAGE BATTERIES

Meanwhile there's an evolution of improved storage batteries, some of which may make electric cars far more popular in just a few years. While these developments and the improvement in performance and lowering of cost of solar cells are occurring in parallel, it's interesting to speculate as to what your future garage and electric commuter car may look like—perhaps as early as 1980. One thing is certain about technological progress: it can take place far more rapidly than conservative scientists and engineers are willing to predict, if there is sufficient popular demand. And as gasoline climbs above $.70 a gallon, there's powerful impetus toward a battery-operated car which you may be able to drive for $.05 a mile or less.

Advantages of Molten Salt Batteries

New improved batteries, using molten salts, provide more power in far less space and with much less weight than conventional lead-acid batteries. General Motors, Ford, Chrysler and American Motors have all built experimental electric cars with these standard batteries. From a cost standpoint, and with half the weight of a car being its battery complement, these early models have had very little appeal. It's almost as if the Big Four were trying to prove that the public doesn't want anything but a gas hog.

Development of More Efficient Battery Cells

Now there are lithium-sulfur and carbon-sodium chloraluminate battery cells which look very promising, as well as quick-charge nickel-cadmium batteries ("ni-cads") and zinc-chloride units. ESB Corporation is doing a lot with batteries having porous carbon plates as the positive electrodes, aluminum plates as negative electrodes, and sodium chloroaluminate as electrolyte. Aluminum ions come off the negative plates, with chloride ions moving to and from the carbon plates.

A system using a brilliant yellow material called chlorine hydrate with zinc plates is being developed by Udylite and tested in a modified Vega Hatchback. There are five battery modules, each containing twelve submodules. A module pump circulates electrolyte through zinc-carbon electrodes in each twenty-four-cell submodule, providing the discharge. When charging this battery, the zinc replates the electrodes, and chlorine is recovered by cooling the electrolyte so that the solid hydrate is formed again. With this battery-operated, modified Vega, a test driver gets 150 miles before needing to recharge the batteries, driving the car at 50 miles per hour (mph). Acceleration is from 0 to 40 mph in ten seconds.

Other programs including new battery designs are featuring sodium-sulfur cells operating at high temperature. Since sodium costs only $.20 a pound and is available in massive quantity from seawater, this entry in the future electric car sweepstakes may be a winner.

A recent announcement by William Ylvisaker, president of Gould, states that his company's new nickel-cadmium battery will permit electric cars to travel eight times as far between charges as cars with lead-acid batteries. A small car like a Vega or Pinto using the new nickel-cadmium battery, says Gould, would weigh about 2,500 pounds (as opposed to its gas-powered weight of about 2,000 pounds), have a top speed of 60 mph, and go for 200 miles between charges. It's beginning to sound better all the time.

ELECTRIC CARS AVAILABLE TODAY

Cars with Conventional Batteries

Citicar

Available for sale now is the two-passenger CitiCar made by Sebring-Vanguard, in Sebring, Florida, operated by batteries of the conventional lead-acid type. (See Figure 77.) There are eight batteries, each 6-volt (v), and the total battery complement weights 524 pounds. There is a built-in battery charger, as well as a 3.5 horsepower GE electric motor and Terrell transmission. There's a variable voltage controller operated from the accelerator so that the voltage starts at 24 v, switches to two parallel 24-v battery banks, and uses 48 v (all eight batteries in series) for maximum speed, which is said to be about 35 mph for cruising. Range between charges is about 50 miles. At present the makers of CitiCar figure that their vehicle can travel 75 miles on electric power generated by one gallon of oil at the utility. Estimated cost of operation is about $.01 per mile. With ordinary maintenance, the battery pack should last for about 15,000 miles or up to 600 charges. Car price is $2,690.

Figure 77. A fleet of battery-operated CitiCars made by Sebring-Vanguard.

Islander Electric Car

A slightly larger and more expensive electric car (four-passenger instead of two) called the Islander has been announced by Electric Vehicle Engineering of Bedford, Massachusetts. It is powered by a 78-volt battery and an 8-horsepower electric motor. This car weighs about 2,200 pounds (nearly double the CitiCar's 1,250 pounds) and sells for $5,000.

Elcar

A foreign entry is Elcar made by Zagate, Milan, Italy. It's lighter than CitiCar, somewhat faster, and with about the same 50-mile range between charges. Price for the newest model is $3,395, including a combination automatic and manual transmission.

Electric Jeeps

Meanwhile, AM General Corporation, a subsidiary of American Motors, is delivering three hundred and fifty electric Jeeps to the United States Postal Service in 1975. This will be the largest fleet of battery-operated cars in this country. Otis Elevator, Cleveland, and Battronic Corporation, Boyertown, Pennsylvania, are building commercial delivery vehicles with battery power.

Future Solar Electric Cars

In the future, with cheaper and more efficient solar cells, you'll be able to roof your garage with some arrays of silicon, gallium arsenide, or other kind of solar cells. Then, says Dr. Mlavsky of Mobil Tyco, you'll have the sun charging your battery modules during the day while you're driving your electric car to work, or shopping, or going to the beach or golf course. At night, you switch battery modules, putting the solar charged module in your car, and connecting the discharged unit to your solar panel.

For longer trips, you'll visit a new kind of service station having a solar cell roof and racks of replacement batteries. You'll be able to rent a suitable battery module, solar-charged, any time your unit needs charging.

Costs of operating your car will dive from $.10 or more per mile at present rates to less than $.05. Maybe it won't be quite as exciting to drive your battery-operated vehicle as that roaring sports car or limousine. But also, when there are enough of us in electric cars, we'll get rid of a major source of most pollution. Imagine being able to see the Milky Way regularly at night in a big city, and to breathe clean air! With some cooperation from us, the sun can help make our living much healthier as well as economical.

Solarex Corporation, Rockville, Maryland, recently announced silicon solar cells with an efficiency of 10% and excellent stability up to temperatures as high as 400°C. This breakthrough is exciting because it proves that, with American ingenuity and money being applied in increasing volume to the problem of making solar cells at 50¢ per watt, this target will be achieved far sooner than the original goal of 1985. The author's guess is, before 1980. Goodbye smog!

12 / Large-Scale Solar Power Plants

GOVERNMENT AND INDUSTRIAL INTEREST IN ALTERNATE ENERGY SOURCES

Like the home owner who's concerned about the steep rise in utility bills, the big electric power companies are worried about the sharp increase in fossil fuel costs. If you're paying 100% more for your electricity than three years ago, remember that your electric utility is paying a comparably greater amount for the fuel burned to generate electricity.

Furthermore, although the demand for electricity will continue to climb as our population increases, utility companies are faced with a major public relations problem when planning new electric plants. If they use fuel oil, gas, or coal to generate the steam for their turbines, then they must satisfy the criteria imposed by environmental planning commissions so as not to harm the ecology of the proposed site and its neighborhood. If a nuclear power plant is contemplated, either at a new site or as an expansion of an existing facility, the problems are far more severe. Many distinguished scientists, including nuclear experts as well as ecologists, are convinced that we should stop building nuclear power plants *anywhere* because of the dangers involved.

For all these reasons, there is increasing interest among utility companies, some sectors of industry, government agencies, and the public, in alternate sources of energy for generating large amounts of electricity. Toward this end, solar energy looks extremely promising. Millions of dollars—which may soon become billions—are being spent to develop this relatively clean, pollution-free source of millions of megawatts. This chapter will examine the most feasible plans.

PROPOSALS FOR LARGE SOLAR POWER PLANTS

Solar Collector Farms

One of the most promising ideas in using solar energy is easy to understand now that you know all about flat-plate solar collectors.

This concept is being tested on a small scale by the inventors, a husband-and-wife scientific team, Drs. Allen and Marjorie Meinel, living in a solar home in Tucson, Arizona.

Dr. Allen Meinel is director of the Optical Sciences Center, University of Arizona, and his wife, Marjorie, is also a distinguished physicist. Their concept of a national solar power facility appears both technically feasible and perhaps economically justifiable.

The Meinels want the government to set aside 5,000 square miles of desert area—some of the badlands you may have seen from the air if you've flown over it—along the Colorado River in California, Nevada and Arizona. This area will then be gradually covered by huge farms of highly efficient solar collector arrays, built on racks and all facing slightly west of south, inclined at 40° from the horizontal.

Each solar collector will include selective black absorbers and will be able to generate very high internal temperatures, up to 1,000° F. The proposed heat exchange medium is liquid sodium, which transfers heat to a large storage tank containing eutectic salts, which serves as a heat reservoir for driving steam turbines.

As an author's note, let me add that the new collectors of the type designed by Corning Glass may be able to achieve such high temperatures. Also, this design, if produced in huge quantities like fluorescent light bulbs, should become relatively inexpensive.

Advantages of Collector-Farms

Fresh water from seawater as a solar plant byproduct

Part of the Meinels' proposal is to pump in huge quantities of seawater from the Gulf of California and/or the Southern California coast. This water would furnish steam for the turbogenerators as well as cool the condensers in a closed-cycle operation. A major benefit would be that part of the electricity developed by the solar system could be used in desalinating the imported salt water. This fresh water would then be used to irrigate neighboring desert areas, and to replenish the Colorado River—benefitting both American and Mexican ranchers.

Solar power is pollution free

If the Meinel program is funded, a 50-megawatt power plant could be in operation in another two years.

Their program for the future is far more ambitious: one billion dollars spent each year over a ten-year period to build a series of twenty big power plants, each producing 1,000 megawatts (Mw), or one billion watts each, from the sun. This program would also vastly increase the total wealth of this area and therefore of the nation. It would eliminate the "need" for building a host of additional nuclear power plants.

Although the dollars for this program may seem large, actually the investment is comparable with other energy sources. The first billion-watt solar plant proposed by the Meinels is estimated to cost $700 million, while subsequent plants would cost less. By comparison, the Navajo power plant in northeastern Arizona cost $616 million and blows coal smoke over previously unpolluted areas. Its annual cost of operation is much higher than that of the proposed solar power plants which use free fuel from the sun.

A further advantage of the Meinel program is that mass production of solar panels for such a huge program would bring the overall cost of collectors down for the individual home owner.

ERDA Solar Tower

A completely different approach has been financed by the Energy Research and Development Adminstration with eight million dollars to four major aerospace companies: Boeing, McDonnell Douglas Astronautics, Martin Marietta, and Minneapolis Honeywell.

In the McDonnell Douglas concept, already tested with company funds on a small scale in the bright high desert of the Naval Weapons Center, China Lake, California, a tower 312 feet high is built. On top of this tower is a boiler on which will be focused sunlight from a large array of mirrors called parabolic heliostats (see Figure 78). Intense solar heat will turn the liquid in this boiler to steam which operates a turbogenerator and produces 10 megawatts of power. This would serve the needs of a small town of ten thousand people.

Some of the basic ideas for this program came from scientists at the University of Houston, who calculated that temperatures up to 1,500° F could be generated in such a solar boiler. It is similar to a proposal made several years ago by a noted Russian, Dr. Valentin A. Baum, director of the Heliotechnical Institute of Moscow. His idea is a series of concentric railroad tracks, each with a locomotive hauling flat cars carrying mirrors to aim sunlight at the boiler on a 131-foot tower. For economic reasons this program was abandoned—too many trains, too many crews, too much fuel consumed for the solar power generated. Hopefully, the field of mir-

Figure 78. Model of a large solar plant being developed by McDonnell Douglas Astronautics Company.

rors in the McDonnell Douglas approach shown in Figure 78 will make dollars and sense as well as megawatts.

Solar Power Plant Using Heated Helium

Completely different is a solar power plant designed by Dr. Howard B. Palmer of Pennsylvania State University, and Dr. Simion C. Kuo, of United Aircraft Research. This is a solar collector-concentrator in which helium is heated in a long glass tube, ⅝ mile long, and comes out at a temperature hot enough to operate a closed-cycle helium turbogenerator. A small experimental model is now in operation in Pennsylvania. If the results are favorable, a solar plant may be built to furnish electricity for a city of 100,000 persons.

There's a slight similarity between the Palmer-Kuo design and the Corning solar collector for home use described in Chapter 3. However, in the big solar power plant, not only will the glass tubes be more than half a mile long, but there will be one hundred of them to combine their output of sun-heated helium. Each pipe is made of high-temperature glass with a narrow window into which is focused solar energy concentrated by an Archimedes mirror. This mirror is actually a series of reflecting slats about 3 inches wide, like a Venetian blind made of mirrors. A control system tilts the mirror surfaces to focus the maximum amount of sunlight through the transparent window in the glass pipe.

Just like your home solar collector, the pipe is carefully insulated and the window area is covered with Tedlar plastic sheet to trap the infrared solar heat. Inside the pipe is an absorber made of a thin slab of black graphite which is heated by the sun and transfers this energy to helium pumped through the pipe. This inert gas enters the pipe's cool end at 160° F and leaves at a temperature seven times as great, about 1,100° F at a pressure of about 115 psia (pounds per square inch absolute). A cross-section is shown in Figure 79.

It's expected that a medium-sized power plant, from 25 Mw to 60 Mw in capacity, can be built using this technique at a cost competitive with conventional designs. Such a design would be most useful in hot dry areas having maximum sunshine all year long. Here again, a useful byproduct might be desalinated seawater. If so, we may have some extremely valuable knowhow and hardware as additional exports to the Organization of Petroleum Exporting Countries (OPEC).

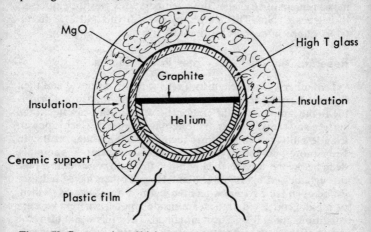

Figure 79. Cross-section of high-temperature glass collector tube by Palmer and Kuo.

Solar Plants in Space

There have been numerous proposals aimed at setting up a vast solar power plant in space, using an inflatable solar collector working in the perpetually clean thin atmosphere available to an orbiting vehicle. Perhaps it would be in synchronous orbit, about 22,000 miles above this planet, and send electrical energy to receivers on the ground by microwave. Return from such an investment looks highly questionable, and hazards at the receiving end could be great.

Power Plants on Tropical Oceans

One of the ways in which this planet stores enormous amounts of solar energy is in the oceans. Particularly in tropical waters, there is a considerable temperature difference between the sun-warmed water at the surface, reaching temperatures of 75° to 80° F, and water deep below, where the temperature goes as low as 35° F at a depth of 2,000 feet.

For nearly two centuries physicists have known that you can transform the difference in temperature between a hot source and a cold one into mechanical work. A principle discovered by Nicolas Carnot says that the maximum efficiency of heat cycle equals (T_1 minus T_2) divided by T_1, where T_1 is the temperature of the hot reservoir and T_2 is the cold source's temperature. For a correct measurement of efficiency these temperatures must be expressed in degrees Kelvin. That simply means that you add 273° to the temperature in degrees Centigrade.

Research on Tropical Water Power Plants

Two other French scientists, first Jacques d'Arsonval and then Georges Claude, decided that you could apply this Carnot principle to the temperature differences between warm seawater at the surface and the cold water in the depths to generate electric power. Claude built a small plant producing 22kw on Matanzas Bay in Cuba more than forty years ago. To get cold water, he ran pipes deep below the surface, then pumped this cold seawater in a pipe through the surf to his power plant. He and others have used a so-called open cycle system where the warm surface water is flashed, or rapidly converted, into steam inside a boiler where a vacuum condition, about 0.5 psi, is maintained. Then this steam turns the vanes of the turbogenerator, and the output is electric power.

There are several groups of modern engineers who have proposed what seems to be a more efficient closed-cycle system for using the solar energy impounded in tropical waters. This involves a Rankine cycle engine. It works very much like the process in a gas-driven or solar-heated refrigeration cycle. The working fluid is propane or ammonia, either of which boils with a high vapor pressure at the relatively low temperatures, like 75° F, which you can get conveniently from seawater. This liquid is first compressed and then goes into an evaporator containing a heat exchanger. Warm seawater flows into this evaporator and causes the liquid—propane, for instance—in the heat exchanger pipes to turn into a vapor under pressure. Now the gas flows into the turbine, expanding through its vanes and driving the electric generator.

To complete the closed-cycle process, the gas from the turbine, which has gone from the high-pressure side to the low-pressure side of the turbine, flows into a condenser. This contains another heat exchanger utilized so that cold seawater from 2,000 feet below the ocean surface causes the gas to return to its liquid state. Next the liquid passes through a pressurizer and is pumped back into the evaporator, where warm seawater again vaporizes the pressurized liquid.

One of the leaders in working with this proposed ocean power plant is J. Hilbert Anderson, former chief engineer of the York Division of Borg-Warner Corporation, and designer not only of advanced air-conditioning and refrigeration equipment but also of the vapor turbines used in the MagmaMax process for obtaining electric power efficiently from geothermal wells.

Anderson and his son, James H. Anderson, Jr., have founded a company called Sea Solar Power in York, Pennsylvania. Other scientists doing similar work include Dr. Abraham Lavi and Dr. Clarence Zener, Carnegie-Mellon University, Pittsburgh, Pennsylvania; and Professor William E. Heronimus, University of Massachusetts, Amherst, Massachusetts.

A proposed sea power plant would be moored in the Gulf Stream about fifteen miles east of Miami, where there's plenty of warm seawater, and it would be possible to bring up water from the depths at a temperature of around 33° F. If federal funding can be obtained, first a pilot plant will be tested offshore to establish feasibility and check the efficiency of various parts of the system such as the heat exchangers.

Then it is considered possible that an electric plant with a capacity of 400 Mw could be built some twenty miles off the Florida coast. Its power would be delivered by submarine cables.

Effects on Marine Ecology

As to the question of whether such power plants would upset the ecology of tropical oceans, scientists think the effects may be more beneficial than harmful. The effect of cycling seawater from the cold depths to the surface, even by numerous sea-solar plants, would be to reduce evaporation by a tiny fraction. This would mean that countries bordering on tropical seas would be slightly less humid and a little cooler.

Also, the cold water brought up from the depths could be used to cultivate shellfish such as oysters and shrimps kept in special pens. By flowing cold water through these pens, the shellfish would be fed useful organic and inorganic materials brought up from two thousand feet.

ERDA has set up a program office in Washington, D.C., under the direction of Dr. Robert Cohen, for monitoring progress toward power from the ocean. If all goes well, in another five years we may be getting megawatts from a novel kind of Texas tower— perhaps leading to many offshore power plants along the Florida coast and the Gulf of Mexico.

THE NEED FOR FUTURE SOLAR POWER PLANTS

No one has yet built a large-scale solar power plant. But with the money now being spent in development, a good part of it from us taxpayers, it is highly probable that there will be electric utilities producing megawatts from solar energy within a few years. Perhaps all we need is political leadership with the vision of the Roosevelt administration that created the Tennessee Valley Authority (TVA). None of the programs described in this chapter takes more funding to get to practical megawatts than was needed to launch the TVA.

As compared with any type of power plant using gas, oil, coal, or nuclear fuel, a solar power plant contributes a tiny fraction of the pollution. This should be a most important consideration for any citizen of this planet, and particularly for Americans. With about 6% of the world's population, we consume nearly 35% of the world's electrical energy. We have managed to pollute urban air, our rivers, lakes and shorelines, and vast areas of strip-mined land.

It's time to take an active interest in developing solar energy, not merely because of the energy crisis and to save ourselves money, but also because we'll be doing something truly constructive for future residents of this earth.

Appendix I

SOLAR MANUFACTURERS

Manufacturers of Liquid-type Solar Collectors

A-Z Solar Products, 200 East 26th Street, Minneapolis, Minnesota 55404

Airtex Corporation, Sidney, Ohio

Albuquerque Western Solar Industries, Inc., Rankin Road NE, Albuquerque, New Mexico 37107

ALCOA, Pittsburgh, Pennsylvania

Alten Associates, Inc., Santa Clara, California

American Building Center, 2626 East Cerritos, Los Alamitos, California 90720

American Heliothermal Corporation, Denver, Colorado

American Solar King Corporation, Waco, Texas

Ametek, Inc., Paoli, Pennsylvania

Applied Sol Tech Inc., P.O. Box 9111 Cabrillo Station, Long Beach, California 90810

Atlantic Solar Products, Reston, Virginia

Berry Solar Products, Edison, New Jersey

Burke Industries, Inc., 2250 South 10th Street, San Jose, California 95112

Butler Ventamatic Corporation, Mineral Wells, Texas

CSI Solar Systems Division, 12400 49th Street North, St. Petersburg, Florida 33732

Calmac Manufacturing Corporation, Englewood, New Jersey

Chamberlain Manufacturing Co., 845 Larch Avenue, Elmhurst, Illinois 60540

Cole Solar Systems, Inc., 440A East St. Elmo Road, Austin, Texas 78745

Columbia Chase Solar Energy Division, 55 High Street, Holbrook, Massachusetts 02343

Daystar Corporation, 90 Cambridge Street, Burlington, Massachusetts 01803

E & K Service Co., 16824 74th Avenue N.E., Bothell, Washington 98011

Edmund Scientific, 101 East Glouster Pike, Barrington, New Jersey 08007

Energex Corporation, 5115 Industrial Road, Las Vegas, Nevada 89118

Energy Converters Inc., 2501 North Orchard Knob Avenue, Chattanooga, Tennessee 37406

Energy Systems Inc., 4570 Alvarado Canyon Road, San Diego, California 92111

Enertech Corp., Golden, Colorado

Fafco, Inc., Menlo Park, California 94025

Flagala Corp., Panama City, Florida

Future Systems, Inc., Lakewood, Colorado

General Electric Solar Division, P.O. Box 8555, Philadelphia, Pennsylvania 19101

Grumman Energy Systems, 4175 Veterans Memorial Highway, Ronkonkoma, New York 11779

Gulf Thermal Corp., Sarasota, Florida

Halstead & Mitchell, Highway 72 West, Scottsboro, Alabama 35768

Harrison Radiator Division, General Motors, A&E Building, Lockport, New York 14094

Heilemann Electric, 127 Mountainview Road, Warren, New Jersey 07060

Helios Corporation, 1313 Belleview Avenue, Charlottesville, Virginia 22901

Heliotherm, Inc., Lenni, Pennsylvania

Hitachi America Ltd., 437 Madison Avenue, New York, New York 10022

Honeywell Inc., Honeywell Plaza, Minneapolis, Minnesota 55408

Intertechnology Solar Corporation, 100 Main Street, Warrenton, Virginia 22186

Kalwall Corporation, Solar Components Division, 1111 Candia Road, Manchester, New Hampshire 03103

Kastek Corporation, Portland, Oregon

Lennox Industries, Inc., P.O. Box 250, Marshalltown, Iowa 50158

Libby-Owens-Ford Technical Center, Toledo, Ohio

National Sun Systems, Inc., St. Petersburg, Florida

Natural Energy Systems, 1001 Connecticut Avenue NW, Washington, D.C. 20036

Northrup, Inc., Hutchins, Texas

Olin Brass, Shamrock Street, East Alton, Illinois 62024

Original Power Equipment Company, Garland, Texas

Owen Enterprises Inc., 436 North Fries Avenue, Wilmington, California 90744

Owens-Illinois Inc., P.O. Box 1035, Toledo, Ohio

PPG Industries Inc., One Gateway Center, Pittsburgh, Pennsylvania 15222

Piper Hydro Inc., 2895 East La Palma, Anaheim, California 92306

Pleiad Industries Inc., West Branch, Iowa

Raypak Inc., 31111 Agoura Road, Westlake Village, California 91361

Refrigeration Research Inc., 525 North Fifth Street, Brighton, Michigan 48116

Revere Copper and Brass Inc., P.O. Box 151, Rome, New York 13440

Reynolds Metals Company, 6601 West Broad Street, Richmond, Virginia 23261

Scientific-Atlanta Inc., 3845 Pleasantdale Road, Atlanta, Georgia 30340

Semco, Solar Products, 1054 N.E. 43rd Street, Fort Lauderdale, Florida 33334

Simons Solar Environmental Systems Inc., Mechanicsburg, Pennsylvania

SolaPlay Inc., Whitewater, Wisconsin

Solar Applications Inc., San Diego, California

Solar Corporation of America, Warrenton, Virginia

Solar Development Inc., 4180 Westroads Drive, West Palm Beach, Florida 33407

Solar Dynamics Inc., Dania, Florida

Solar Energy Company, Merrimack, New Hampshire

Solar Energy Products Inc., Gainesville, Florida 32601

Solar Energy Systems, Carson, California

Solar Energy Systems of Georgia, 2616 Old Wesley Chapel Road, Decatur, Georgia 30339

Solar Energytics Inc., Jasper, Indiana

Solar Equipment Sales Co., Scarsdale, New York

Solar-Eye Products, Fort Lauderdale, Florida

Solar Farms, Stockton, Kansas

Solar Industries Inc., Monmouth Airport Industrial Park, Farmingdale, New Jersey 07727

Solar Living Inc., P.O. Box 15345, Austin, Texas 78761

Solar Research Systems, Costa Mesa, California 92626

Solar Shelter, Reading, Pennsylvania

Solar Shingle, Straza Enterprises, 1071 Industrial Place, El Cajon, California 92020

Solar Systems Sales, Novato, California

Solar Sun Inc., 235 North 12th Street, Cincinnati, Ohio 45210

Solar II Enterprises, Los Gatos, California

Solarator Inc., Madison Heights, Michigan

Solaray Corporation, Honolulu, Hawaii

Solarcoa Inc., 2115 East Spring Street, Long Beach, California 90806

Solarmaster, Santa Maria, California

Solartec Corporation, 8250 Vickers, San Diego, California 92111

Sol-R-Tech Inc., The Trade Center, Hartford, Vermont 05047

Solus Inc., Houston, Texas

Sol-Therm Corp., 7 West 14th Street, New York, New York 10011

Southeastern Solar Systems, Inc., 4705J Bakers Ferry Road, Atlanta, Georgia 30336

Southern Lighting Mfg., Orlando, Florida

Southwest Ener-Tech, 3030 Valley View Boulevard, Las Vegas, Nevada 89102

State Industries Inc., Ashland City, Terrace 37015 and Henderson, Nevada 89015

Sun Century Systems, Florence, Alabama

Sun Power Systems, Ltd., Sunnyvale, California

Sun Source Inc., 9570 West Pico Blvd., Los Angeles, California 90035

Sun Sponge, 1288 Fayette Street, El Cajon, California 92020

Sun Systems, Inc., Eureka, Illinois

Sunburst Solar Heating, Inc., Menlo Park, California 94025

Sundu Company, 3319 Keys Lane, Anaheim, California 92804

Sunearth Inc., Montgomeryville, Pennsylvania

Sunsav Inc., Lawrence, Massachusetts

Suntap Inc., 42 East Dudley Town Road, Bloomfield, Connecticut 06002

Sunwater Company Inc., 1654 Pioneer Way, El Cajon, California 92020
Sunworks Division of Enthone Inc., P.O. Box 1004, New Haven, Connecticut 06508
Tranter, Inc., 735 East Hazel Street, Lansing, Michigan
Tri-State Solar King Inc., Adams, Oklahoma
United States Solar Systems Inc., Los Angeles, California
Wallace Company, Gainesville, Georgia
Ying Manufacturing Corp., 1957 West 144th Street, Gardena, California 90249
ZZ Corporation, 10806 Kaylor Street, Los Alamitos, California 90720
Zomeworks, 1212 Edith N.W., Albuquerque, New Mexico 87125

Manufacturers of Air-type Solar Collectors

American Heliothermal Corporation, Denver, Colorado
Champion Home Builders Co., 5573 East North, Dryden, Michigan 48428
Future Systems, Denver, Colorado
Kalwall Corporation, Solar Components Division, P.O. Box 237, Manchester, New Hampshire 03105
NRG, Napoleon, Ohio
Solar Inc., P.O. Box 246, Mead, Nebraska 68041
Solar Store, Parker, South Dakota
Solaron Corporation, 4850 Olive Street, Commerce City, Colorado 80022
Sun Unlimited Research Corporation, P.O. Box 941, Sheboygan, Wisconsin 53081
Sunworks Division of Enthone Inc., P.O. Box 1004, New Haven, Connecticut 06508
Ying Manufacturing Corp., 1957 West 144th St. Gardena, California 90249

Appendix 2

NUMBER OF GARDEN WAY PANELS REQUIRED FOR DOMESTIC HOT WATER PER PERSON

City	Approximate Latitude	Summer Panel Requirement (per person)	Winter Panel Requirement (per person)
Albuquerque, New Mexico	(35°)	0.5	1.4
Annette Island, Alaska	(55°)	1.2	4.0
Apalachicola, Florida	(30°)	0.7	1.2
Astoria, Oregon	(46°)	0.9	2.8
Atlanta, Georgia	(33°)	0.7	1.8
Barrow, Alaska	(71°)	1.3	175.4
Bethel, Alaska	(60°)	1.2	11.2
Bismarck, North Dakota	(47°)	0.7	2.9
Blue Hill, Massachusetts	(42°)	0.9	2.9
Boise, Idaho	(43°)	0.6	2.9
Boston, Massachusetts	(42°)	0.8	2.9
Brownsville, Texas	(26°)	0.6	1.2
Caribou, Maine	(47°)	0.9	4.0
Charleston, South Carolina	(33°)	0.7	1.5
Cleveland, Ohio	(41°)	0.8	3.2
Columbia, Missouri	(39°)	0.7	2.3
Columbus, Ohio	(41°)	0.7	3.1
Davis, California	(38°)	0.6	2.1
Dodge City, Kansas	(38°)	0.6	1.6
East Lansing, Michigan	(42°)	0.8	3.7
East Wareham, Massachusetts	(42°)	0.9	2.4
Edmonton, Alberta	(53°)	0.9	4.6
El Paso, Texas	(32°)	0.5	1.3
Ely, Nevada	(39°)	0.6	1.8
Fairbanks, Alaska	(65°)	0.9	83.3
Fort Worth, Texas	(33°)	0.6	1.4
Fresno, California	(37°)	0.6	1.9
Gainesville, Florida	(29°)	0.7	1.2
Glasgow, Montana	(48°)	0.6	3.4

City	Approximate Latitude	Summer Panel Requirement (per person)	Winter Panel Requirement (per person)
Grand Junction, Colorado	(39°)	0.6	1.8
Grand Lake, Colorado	(40°)	0.8	2.6
Great Falls, Montana	(47°)	0.7	2.8
Greensboro, North Carolina	(36°)	0.7	1.9
Griffin, Georgia	(33°)	0.7	1.7
Hatteras, North Carolina	(35°)	0.6	1.5
Indianapolis, Indiana	(40°)	0.7	2.9
Inyokern, California	(35°)	0.5	1.1
Ithaca, New York	(42°)	0.8	3.6
Lake Charles, Louisiana	(30°)	0.7	1.4
Lander, Wyoming	(30°)	0.7	2.2
Las Vegas, Nevada	(36°)	0.5	1.2
Lemont, Illinois	(41°)	0.8	2.6
Lexington, Kentucky	(38°)	0.7	2.0
Lincoln, Nebraska	(41°)	0.8	2.1
Little Rock, Arkansas	(34°)	0.7	1.8
Los Angeles, California	(34°)	0.7	1.2
Madison, Wisconsin	(43°)	0.8	2.8
Matanuska, Alaska	(61°)	1.1	14.2
Medford, Oregon	(42°)	0.6	3.1
Miami, Florida	(26°)	0.7	0.9
Midland, Texas	(32°)	0.6	1.6
Nashville, Tennessee	(36°)	0.7	2.1
Newport, Rhode Island	(41°)	0.8	2.4
New York, New York	(41°)	0.8	2.5
Oak Ridge, Tennessee	(36°)	0.7	2.1
Oklahoma City, Oklahoma	(35°)	0.6	1.4
Ottawa, Ontario	(45°)	0.8	3.6
Phoenix, Arizona	(33°)	0.5	1.2
Portland, Maine	(43°)	0.8	2.3
Rapid City, South Dakota	(44°)	0.7	2.2
Riverside, California	(34°)	0.6	1.2
St. Cloud, Minnesota	(45°)	0.7	3.4
Salt Lake City, Utah	(41°)	no data	2.3
San Antonio, Texas	(29°)	0.6	1.3
Santa Maria, California	(35°)	0.7	1.4
Sault Ste. Marie, Michigan	(46°)	0.8	3.8
Sayville, New York	(40°)	0.8	2.4
Schenectady, New York	(43°)	0.9	3.7
Seattle, Washington	(47°)	0.9	3.7
Seabrook, New Jersey	(39°)	0.8	2.4

City	Approximate Latitude	Summer Panel Requirement (per person)	Winter Panel Requirement (per person)
Spokane, Washington	(47°)	0.6	4.0
State College, Pennsylvania	(41°)	0.8	3.1
Stillwater, Oklahoma	(36°)	0.7	1.6
Tampa, Florida	(28°)	0.7	1.0
Toronto, Ontario	(43°)	0.8	3.6
Tucson, Arizona	(32°)	0.6	1.1
Upton, New York	(41°)	0.8	2.2
Washington, D.C.	(39°)	0.7	2.1
Winnipeg, Manitoba	(50°)	0.8	5.0

Conversion of Pitch To Degrees

Angle in Degrees	Pitch in Inches Per Foot
10°	2.1
15°	3.2
20°	4.4
25°	5.6
30°	6.9
35°	8.4
40°	10.1
45°	12.0

Correction Factors for Panel Requirements

Roof Pitch	Multiply by	
	Summer	Winter
Latitude minus 20°	0.97	1.20
Latitude minus 10°	1.00	1.00
Latitude	1.08	0.90
Latitude plus 10°	1.20	0.85
Latitude plus 20°	1.39	0.82

Appendix 3

FACTS ABOUT THE SUN

Most of these statistics for the sun represent the best approximations based on available scientific data and calculations.

Solar power delivered to the earth daily . . . . 85 trillion kilowatts
Note: The sun delivers enough energy in 15 minutes each day to supply the world's total energy needs for a year.
Solar energy per square foot
in the United States . . . . 75 to 104 Btu per hour in summer
21 to 46 Btu per hour in winter
Distance of sun from earth . . . . 92,913,000 miles
Difference in distance of sun from earth
between January and June . . . . 3,069,000 miles
Diameter of sun . . . . 864,000 miles
Note: The sun is almost 110 times greater in diameter than the earth.
Volume of sun compared to earth . . . . 1,3000,000 : 1
Mass of sun compared to earth . . . . 332,488 : 1
Surface gravity of sun compared to earth . . . . 28 : 1
Note: 1 "solar g" equals 28 g on earth
Age of sun from radioactive dating . . . . 5 billion years
Orbital velocity of earth around sun . . . . 18.5 miles per second
Surface temperature of sun . . . . 10,800° F (6,000° C)
Interior temperature of sun . . . . 27,000,000° F
(15,000,000° C)
Wavelength of sunlight . . . . 0.3 to 3.0 microns
(1 micron = 1 millionth of a meter)
Note: This includes invisible ultra-violet solar radiation, the shortest wavelengths in the solar spectrum, and equally invisible infra-red long waves.

Glossary of Technical Terms

(For terms particular to the world of solar energy, see pages 9-10.)

Ambient. The surrounding atmosphere. Thus, **ambient temperature** means the temperature of the atmosphere at a particular location.

Automatic damper. A device which cuts off the flow of hot or cold air to or from a room when the thermostat indicates that the room is warm enough or cool enough.

Azimuth. The angle of horizontal deviation, measured clockwise, of a bearing from a standard direction, as from north or south.

British thermal unit (Btu). The quantity of heat required to raise the temperature of one pound of water one degree Fahrenheit at 39.2° F.

Bushings. Removable cylindrical linings for an opening, such as the end of a pipe. These linings limit the size of the opening, resist abrasion, and act as a watertight guide. Thus, when two different sizes of pipe are mated, one or more bushings might be used.

Carriage bolt. (See **Coach screw**.)

Check valve. Used to turn off the water going through a manually controlled hot water system (or pool).

Coach screw. Like a carriage bolt, it is usually a wood screw with a round head and a square shank above the screw threads. The square shank is useful for a firm fastening as it penetrates the wood.

Conduction. The transfer of heat by contact with a hot body.

Connecting socket (for pipe). A holder of metal or plastic, which is usually threaded to mate with one or more pieces of pipe. A section of pipe may have a male thread at one end, and this is attached to a connecting socket having a female thread of the same pitch.

Convection. The transfer of heat energy by means of the circulation of a liquid or gas.

Degrees Kelvin. A thermometric scale in which the unit of measurement equals one degree centigrade but the initial point is different. That is, the Kelvin scale is so constructed that absolute zero corresponds to $-273.16°$ C. Hence 0° C (the freezing point of water at sea level) is equal to $+273°$ K.

Dessicant-type spacers. Small objects which hold two sheets of glass, or other materials, apart. Dessicant-type spacers include a water-absorbing material such as calcium chloride to keep condensation from forming on the glass plates. Condensation in a collector panel would keep out some of the sun's energy.

Edge retaining system. The steel channel which holds in place the edges of the various layers of a collector panel.

Electroplating. The process of putting a metallic coating or plating on a base material, which is usually metal or plastic, by means of electrodeposition. That is, the metallic plating is deposited on the base by passing an electric current through a plating bath in which are dipped the objects to be plated.

Emissivity. The relative power of a surface to emit heat by radiation, or the ratio of the radiant energy emitted by a surface to that emitted by a black body—considered to have almost perfect heat absorption and therefore very low emissivity—when the given surface and the black body are at the same temperature.

Eutectic material. A chemical which has the property of changing from a solid to a liquid at a relatively low temperature while maintaining a constant temperature. The eutectic liquid then stores the heat energy which caused the transformation until the liquid returns to solid form and gives up heat. Eutectic salts, stored in plastic tubes, are used as thermal reservoirs to conserve and then release solar heat.

Flashing. Sheet metal used in waterproofing roof valleys or ridges.

Gasket. A piece of solid material, usually metal, rubber, plastic or asbestos, placed between two pieces of pipe or between an automotive cylinder head and the cylinder block, for example, to make the metal-to-metal structure fluid-tight.

Gate valve. Used to regulate the flow of the main supply of water.

Grouting. The filling or finishing of cracks with a thin mortar or caulking compound or any similar material which will mend the cracks and make the surface water-resistant.

Header or **header pipe.** A section of pipe which carries the main liquid flow at the top and the bottom of a solar collector panel, and from which smaller pipes or channels extend to the panel. The bottom header carries cool water to the panel(s). The top header carries sun-heated water away from it. A header is usually at least 1½ inches in diameter and may be 2-inch pipe.

Heat sink. A structure designed to absorb excess heat. In electrical or electronic equipment, a heat sink is often used in a power supply where such components as power transistors and transformers tend to get considerably hotter than the ambient or surrounding atmosphere. These hot components are mounted on a heat sink, which is usually designed with cooling fins to carry away the excess heat by convection.

Hertz (Hz). A unit of frequency used in electrical and electronic measurements and equal to one cycle or one wavelength of electrical energy per second.

Hose coupler. A metal or plastic device for joining two pieces of hose (or one hose and a piece of rigid piping) by means of screw threads in the coupler which mate with threaded connections on the hoses or pipes.

Hysteresis. The time lag exhibited by a body in reacting to changes in the forces affecting it.

Immersion-formed layer. A layer formed by immersion of a solid in a liquid, with the layer formation usually occurring because of electrolytic action.

Incident. In reference to solar radiation, those light rays falling on or striking a surface.

Manifold. See header.

Photovoltaic. The ability of a material to generate electrical energy when it is exposed to radiant energy such as sunshine.

Pinch valve. A valve used to pinch off, or stop, the flow of liquid, usually water, at a desired location.

Pounds per square inch absolute (psia). The measurement of pressure, whether hydraulic (liquids) or pneumatic (gases), made without including the effect of atmospheric pressure, which is 15 psi at sea level.

Rankine cycle. The complete expansion cycle of a gas which is brought into a structure like a turbine engine, where the gas is compressed and then expands rapidly to give off energy.

Rectifier. The container or chamber in a cooling system where, typically, ammonia vapor is separated from water vapor.

Reflective loss. The energy which strikes a surface and is not absorbed but reflected from it.

Saddle fitting. A clamp which is attached over a piece of pipe like a saddle on a horse.

Selective black paint. More absorbent of the infrared long wavelengths of sunlight than non-selective black paint, hence, an improved material for coating the absorber plates in solar collectors.

Solenoid. An electromechanical device so designed that when electric current is applied to a coil of wire wound around a cylinder, the electromotive force causes a bar or plunger inside the cylinder to move.

Ton of cooling. 12,000 Btu per hour. The term is derived from the amount of heat energy required to convert a ton of water into ice at 320° F during a 24-hour period.

Turnbuckle. A device consisting of a link with screw threads at each end. This link is turned to bring the two objects connected by the turnbuckle closer together.

Two-outlet heater. A water heater having one outlet going typically to the piping for a domestic hot water system and the other outlet, through a pipe to a large storage tank.

Vacuum-evaporated film. A film generally formed on a sheet or plate by electrical evaporation of a metal or alloy in an evacuated chamber.

Bibliography

American Institute of Architects Research Corporation. 1975. *Innovation in Solar Thermal House Design*. Prepared by Donald Watson (Guilford, Conn.). Washington, D.C.

——. 1975. *Solar Energy: An Introduction*. Washington, D.C.

——. 1975. *Solar Energy and Housing*. Prepared by Giffels Associates, Inc. (Detroit). Washington, D.C.

——. 1975. *Solar Energy Housing Design*. Prepared by Total Environmental Action (Harrisville, N.H.). Washington, D.C.

——. 1975. *Solar Heated Houses for New England*. Prepared by Massdesign (Cambridge, Mass.). Washington, D.C.

——. 1975. *Solar Heating and Cooling Demonstration Act Information Packet*. Washington, D.C.

——. 1975. *Solar-Oriented Architecture*. Washington, D.C.

——. 1975. *The Design of Solar Heated and Cooled Dwellings*. Washington, D.C.

Arctander, Erik H. January 1975. Solar-MEC: one box heats and cools your house. *Popular Science*.

Brown, Warner. May 1975. Battery-run car coming to power. *Moneysworth*.

CitiCar is no racer, but it moves people. *Christian Science Monitor*. May 1975.

Cobb, Hubbard H. June 1975. Solar heat—your way out of high heating costs? *Woman's Day*.

Daniels, Farrington. 1964. *Direct Use of the Sun's Energy*. New Haven: Yale University Press.

de Winter, Francis. 1974. *Solar Energy and the Flat Plate Collector: An Annotated Bibliography*. New York: Copper Development Association.

Diamant, R. M. E. 1970. *Total Energy*. Elmsford, New York: Pergamon Press.

Edmondson, William B. 1974. *SolarSan Solar Water Heaters and Their Application*. San Diego, Ca.

Edmondson, William B., ed. 1972-75. *Solar Energy Digest*. San Diego, Ca.

Eibling, James A., and Frieling, Donald H. 1974. *A Solar Heat Pump*. Columbus, Oh.: Battelle Columbus Laboratories.

Environmental teaching tool, an. *Princeton Alumni Weekly*. March 18, 1975.

Fafco, Inc. 1975. *Swimming Pool Solar Heater Installation Manual*. Menlo Park, Ca.

Fred Rice Productions. 1973. *The Solar/Sonic Home*. Van Nuys, Ca.

Free, John R. October 1974. To help relieve the energy crunch . . . super-batteries. *Popular Science*.

Fisher, Arthur. June 1975. Energy from the sea. Part II: Tapping the reservoir of solar heat. *Popular Science*.

——. December 1974. Solar cells: when will you plug into electricity from sunshine? *Popular Science*.

Garden Way Laboratories. 1975. *Sol-R-Tech Operations Manual*. Charlotte, Vt.

——. 1975. *Planning Guide for the Garden Way Solar Heat Collector System*. Charlotte, Vt.

Gilmore, C. P. March 1974. Can sunshine heat (and cool) your house? *Popular Science*.

Halacy, D. S., Jr. 1973. *The Coming Age of Solar Energy*. rev. ed. New York: Harper and Row.

Hitachi America, Ltd. 1975. *Hitachi Solar Pre-Heat System*. New York.

Hottel, H. C., and Woertz, B. B. 1940. *The Performance of Flat-Plate Solar-Heat Collectors*. Solar Energy Research Project, publication No. 3. Cambridge, Mass.: Massachusetts Institute of Technology.

International Solarthermics Corp. 1975. *Solar Furnace Specifications and Performance Handbook*. Nederland, Colo.

James, L. W., and Moon, R. L. 1975. *GaAs Concentrator Solar Cells*. Palo Alto, Ca.: Varian Associates.

Keyes, John H. 1975. *Harnessing the Sun*. Dobbs Ferry, N.Y.: Morgan and Morgan.

Lindsley, E. F. March 1975. Storable, renewable hydrogen power: key to unlocking energy from the sun, wind, tides. *Popular Science*.

Mantell, Charles L. 1970. *Batteries and Energy Systems*. New York: McGraw-Hill.

Mlavsky, Alexander I. 1974. *The Silicon Ribbon Solar Cell—A Way to Harness Solar Energy*. Waltham, Mass.: Mobil Tyco Solar Energy Corp.

Mog, Dennis, D. 1975. *Solar Energy: Older than the Earth Itself*. Corning, N.Y.: Corning Glass Works.

——. 1975. *Tubular Evacuated Solar Collectors*. Corning, N.Y.: Corning Glass Works.

Omnium-G. 1975. *Solar Generating System*. Anaheim, Ca.

Palmer, Howard B., and Kuo, Simion C. 1974. Solar farms utilizing low-pressure closed-cycle gas turbines. Unpublished paper prepared for United Aircraft Research Laboratories, East Hartford, Conn.

Payne, Jack, and McConaghie, Tom. September 1974. Solar water heater for your vacation home. *Popular Science*.

Piper Hydro, Inc. 1975. *Conserve In Comfort*. Anaheim, Ca.

Power-test home gets 4 charges. *Combined News Services*. March 1975.

PPG Industries. 1974. *Baseline Solar Collector*. Pittsburgh.

Rau, Hans. 1964. *Solar Energy*. Edited and revised by D. J. Duffin. New York: MacMillan.

Revere Copper and Brass, Inc. 1975. *Revere Solar Energy Collector*. Rome, N.Y.

Schlesinger, Robert J. 1974. *Differential Thermostat Application Notes*. Tarzana, Ca.: Rho Sigma.

Schuessler, Raymond. April 1975. Oceans: power plants in disguise. *Passages* (Northwest Orient Airlines).

Shuldiner, Herbert. June 1975. Battery-powered cars you can buy now. *Popular Science*.

Shurcliff, W. A. 1975. *Solar Heated Buildings: A Brief Survey*. 8th ed. Cambridge, Mass.

Solar Water Heater Co. 1974. Plans for *Solar House and Water Heating*. Coral Gables, Fla.

Solaron Corp. 1974. *Solar Heating and Cooling Now*. Denver.

Stepler, Richard. March 1975. Now you can buy solar heating equipment for your home. *Popular Science*.

Sunsource, Inc. 1974. *The Residential Solar Energy System*. Beverly Hills, Ca.

Sunworks, Inc. 1975. *Design Criteria for Solar-Heated Buildings*. Guilford, Conn.

Thomason, Harry E., and Thomason, Harry Lee, Jr. 1975. *Solar House Plans: Solar House Heating and Air-Conditioning Systems, Solar Greenhouse and Swimming Pool*. Barrington, N.J.: Edmund Scientific Co.

United Nations. 1964. *United Nations Conference on New Sources of Energy*, vols. 4-6. New York.

Watson, Donald. March-April 1974. Energy conservation in architecture, part 1: adapting design to climate. *Connecticut Architect*.

Watson, Donald, and Barber, Everett, Jr. May-June 1974. Energy conservation in architecture, part 2: alternative energy sources. *Connecticut Architect*.

Index

Page numbers for illustrations are printed in italics.

233

Mentor and Signet Books you'll want to read

☐ **OIL POWER: The Rise and Imminent Fall of an American Empire** by Carl Solberg. The great American oil companies and their influence on our country's social, political, economic and military structures . . . "A gripping narrative fact-filled . . . persuasive."—*Publishers Weekly* (#MJ1531—$1.95)

☐ **FROM KNOW-HOW TO NOWHERE: The Development of American Technology** by Elting E. Morison. An exploration of the ways in which machines have been shaped by and in turn shaped American society over the past two hundred years. (#MJ1539—$1.95)

☐ **HOW DOES IT WORK?** by Richard M. Koff. From air conditioners to zippers understand the gadgets and machines in your life—for all emergencies, for knowledge, for fun. Includes more than 300 illustrations. (#W6920—$1.50)

☐ **WHAT YOU SHOULD KNOW BEFORE YOU HAVE YOUR CAR REPAIRED** by Anthony Till. Get the facts on the Great American Repair Racket! This book can save you big money the next time you need any kind of work done on your car. It includes a special "Flat-Rate Book" to show how much you should pay for all major and minor car repairs jobs. (#N5087—$1.00)

☐ **CONSUMER GUIDE: CAMPING AND BACKPACKING** by the editors of CONSUMER GUIDE Magazine. A complete guide to camping and the latest backpacking gear. Includes money-saving ideas on buying tents, sleeping bags, backpacks and frames, stoves, and other equipment to make camp life easier, safer, and more convenient. (#J7082—$1.95)

Quality Non-fiction from Meridian and Plume

☐ **TECHNOLOGY AND CULTURE:** An Anthology edited by Melvin Kranzberg and William H. Davenport. Are machines our servants—or are our lives shaped by our inventions? In this anthology leading authorities examine the vital question of the role of the machine in human history and human destiny.
(# F426—$4.95)

☐ **RECIPES FOR HOME REPAIR** by Alvin Ubell and Sam Bittman. At your fingertips: all the ingredients, instructions, and illustrations that make home and apartment repairs easy. A Book-of-the-Month Club Alternate Selection. (# Z5125—$2.95)

☐ **THE NATURAL HOUSE** by Frank Lloyd Wright. Here, shown in photographs, plans, and drawings, are houses for people of limited means, each individually designed to fit its surroundings and to satisfy the needs and desires of its owners.
(# F445—$3.95)

☐ **THE LIVING CITY** by Frank Lloyd Wright. Mr. Wright unfolds his revolutionary idea for a city of the future, a brilliant solution to the ills of urbanization whereby man can attain dignity in his home, his work, his community. Includes Wright's amazing plans for his model community, Broadacre City.
(# F444—$3.95)

☐ **THE FUTURE OF ARCHITECTURE** by Frank Lloyd Wright. In this volume the master architect looks back over his career and explains his aims, his ideas, his art. Also included is a definition of the Language of Organic Architecture as the architect has employed it throughout a lifetime of work.
(# F446—$3.95)

Ecology Books from Mentor and Signet

☐ **THE WEB OF LIFE** by John H. Storer. This exciting, easy-to-understand introduction to the science of ecology shows how all living things—from bacteria to men—fit into a pattern of life and depend upon each other and the world around them for existence. (# MY1497—$1.25)

☐ **THE SEA AROUND US** by Rachel L. Carson. This outstanding best-seller and National Book Award winner gives an enthralling account of the ocean, its geography and its inhabitants. (# MJ1594—$1.95)

☐ **THE EDGE OF THE SEA** by Rachel L. Carson. A guide to the fascinating creatures who inhabit the mysterious world where sea and shore meet—from Maine's rocky coast to the coral reefs beyond the Florida Keys. Illustrated. (# Q4368—95¢)

☐ **KING SOLOMON'S RING** by Konrad Z. Lorenz. A modern classic on animal behavior by a scientist whose gifts for storytelling and communicating with animals have made him famous. "Charm, lightness . . . carries the reader along with sustained and fascinated attention."—*The New York Times* (# E7816—$1.75)

☐ **THE BIOLOGICAL TIME BOMB** by Gordon Rattray Taylor. The author discusses the new discoveries for the manipulation of life which are being made in biology, supplies a wealth of social problems that will occur and submits that things have gone too far too quickly. (# MW1457—$1.50)